AF810905

Think about it. They lived, worked and fought from a fortress of metal, breathing jet fumes and sucking oxygen from rubber hoses. They were responsible for keeping their planes and their fellow crewmembers in the air, responsible for hitting the targets they were assigned, just responsible. If the crews came back from a combat mission, there might be no thanks at all, but if they didn't come back ... well, they still live on in the memories of their friends and family.

No one ever made a movie about them personally, just them; at least not one with a Hollywood leading man. But for the rest of their lives, the men who called themselves "Crewdogs" would know that their time spent in the air and on that plane would count as some of the most important times of their lives.

In this collection, told in the words of these men, Tommy Towery does what he has always done in his writing -- he shows us the value of ordinary people, wrapped in extraordinary circumstances.

How can such stories not be told?

– Rick Bragg

A Pulitzer Prize winning journalist and bestselling author, Rick Bragg is the gifted author of two best selling books, "*Ava's Man*" and "*All Over but the Shoutin'*." More recently he has used his unique story telling writing skills in *"I am a Soldier, Too – The Jessica Lynch Story."*

The
B-52
Collection

We Were Crewdogs

I

Edited by
Tommy Towery

This book was produced as a cooperative, not-for-profit venture of the contributing authors. It was originally published without the Vol. I as part of the title. With the release of Vol. II, this printing includes the Vol. I on the cover for clarity. Should we decide to publish another volume of stories such as this and you might want to participate, then please contact the editor to let your desire to participate be known or visit our web site at:

www.wewerecrewdogs.com

Contact Info:
Tommy Towery
5709 Pecan Trace
Memphis, TN 38135
ttowery@memphis.edu

ISBN 1-59872-294-8
First Printing, November 2005
Second Printing, January 2006
Third Printing, November 2006
Printed by Instantpublisher

Contributing Authors

Bill Beavers
Lt. Col. USAF, Ret.
B-52 Radar Navigator

Walter J. Boyne
Col. USAF, Ret.
B-52 Pilot

Charles N. Brown
Lt. Col. USAF, Ret.
B-52 Navigator

Robert G. Certain
Col. USAF
B-52 Navigator

George Robert Dempsey
Former Capt. USAF
B-52 Navigator

George W. Golding
Col. USAF, Ret.
B-52 Pilot

Gordon Hallgren
Maj. USAF, Ret.
B-52 EWO

Gary E. Henley
Col. USAF, Ret.
B-52 EWO

James Hooppaw
Col. USAF, Ret.
B-52 Pilot

Arthur Craig Mizner
Maj. USAF, Ret.
B-52 Pilot

Karl Nedela
SMsgt. USAF, Ret.
B-52 Gunner

Vince Osborne
Col. USAF, Ret.
B-52 EWO

Denny Scruggs
Col. USAF, Ret.
B-52 Pilot

Wayne Shimet
Lt. Col. USAF, Ret.
B-52 EWO

Tommy Towery
Maj. USAF, Ret.
B-52 EWO

Lynn Wakefield
Col. USAF, Ret.
B-52 EWO

Gerald Wickline
Col. USAF, Ret.
B-52 Pilot

Table of Contents

Preface

Mission

The Road to Becoming A BUFF Crewdog

Peacetime SAC Warriors

Guarding Our Freedom -The Cold War

The B-52 in Combat -The War in Vietnam

Unclassified Materials

Mission Debrief

Our Mission

Alert morning briefing, Carswell AFB, TX

"We Were Crewdogs" - Mission Statement

Tommy Towery

"Our mission is to fly and fight, and don't you forget it!" That was the motto by which most Crewdogs lived in their service to the United States Air Force. A less known saying (much less formal, but just as good) was put forth by a commander under which I served in midstream of my military service. He had a sign made that hung over his desk that said, "Tell the Crewdogs what you want done, then get the Hell out of the way!"

As an adult I realized that when I was a child I had no idea of what my father had done when he served in the Army in World War II. Had it not been for the fact that he was missing a leg because he had stepped on a land mine at Omaha Beach on D-Day, then I probably would know even less about him. I missed every opportunity I was given to find out more about his life, and when he died, the story of his military service in World War II went with him to the grave. To me it is sad that the story of his sacrifices for the freedom, which I and my fellow Americans now enjoy, will never be shared with the next generation.

I do not want that to happen to my generation. My daughter was born after the Vietnam conflict had ended. Except for the odd stories she has overheard between others and me who were there, she has little concept of what I did in B-52s, just as I was unaware of my own father's World War II life. She has even less understanding of the rest of my B-52 experiences, including the approximate 120 days a year I spent on nuclear alert during the early years of her life. Long after my memory has failed me, when her children ask what their grandfather did in the Air Force, I want something documented to explain it to them.

I know that many stories have been written about the combat missions flown in Vietnam. That was the conflict in which most of my generation participated and the one with which they most identify. On the other hand, there have been very few things written about the day-to-day training, peacetime missions, and everyday life of a B-52 Crewdog during the Cold War era. When the Strategic Air Command ended its Nuclear Alert readiness in 1991, I was already out of the B-52 and retired from the service. Now it seems even longer in the past, and not only do aircrews on Nuclear Alert no longer exist, neither does the Strategic Air Command, the command in which I spent the majority of my 20 years of military service.

That's the educational goal of this collection of stories. Some people sit down and write stories to educate others. Some people write stories to entertain. Some record history. We hope that in our endeavors, this group of stories will do all of these things. It is a collection of the serious and the humorous sides of living the life of a SAC B-52 Crewdog, and it is intended to share those experiences with others. The common thread that binds them is the B-52 aircraft, the BUFF (Big Ugly Fat Fellow).

Basically, as editor, I told the Crewdogs what I wanted and then got the Hell out of the way. Each author was encouraged to select the subject content and writing style he wanted to use to tell his part of the story and to write it in his own personal words. Once the stories were completed, an attempt was made to put them in some logical order. Some were easy to classify but others seemed to fit no particular category, so the best match available was selected. These accounts came from officers and enlisted men alike. They are serious and humorous and represent similar stories remembered by many ranks. At least one member of each crew position of the B-52 contributed to this collection. There are many more similar stories that will never be recorded.

As a bonus, we have included several pages in the front of this book to help individuals record their own personal B-52 Crewdog information and hope that they prove helpful for others who might read this book.

My B-52 Crewdog Experience

Place Your Photo Here

And Record

Your Personal

Information In The

Spaces Below

Name

Crew Position

Flight Training

Memorable Crews and Crewmembers

B-52 Bases Assigned, Squadrons, and Dates

B-52 Models Flown

Flight Hours (i.e. Combat, B-52, Total)

Combat Missions Information

Memorable TDY deployments

Decorations Awarded

Special Recognitions

Additional

Photo

Personal Memories

Personal Memories

The Road To

Becoming A

BUFF Crewdog

Lt. George Robert Dempsey, in his 1505 khaki uniform at
Mather AFB, CA shortly after graduation from Navigator's Training

The Language of the B-52 Crewdogs

To understand the stories, you have to know what they're saying. Most books put the glossary in the back, but this one is located in the front. It is done this way to help you better comprehend many of the stories as you read them. As in every profession, many words and phrases make their way into the everyday conversations of the participants. They can talk for hours in a crowded room, and most of their conversation might as well be in Ancient Greek to those who overhear these dialogs. As in most other professions, Crewdogs converse with these terms without any confusion as to what they were saying or meaning. Many wives and dependants were also exposed to this strange language enough that they could understand their meanings as well.

You will find that there are many words and their spellings that were and will forever, it seems, remain sources of debate among many fliers. For many of them, the preferred spelling is "Crew Dog." For others, it is all one word "Crewdog." The same goes for the nickname given to the B-52. Is it "BUF" or "BUFF"? Is "Vietnam" or "Viet Nam" correct? How about "U-Tapao" or "Utapao"? While there is an attempt to standardize the spellings in this collection, some of the stories might use one or the other versions, depending upon who does the writing. In the long run, it doesn't really matter, they all mean the same and in the spoken language they sound the same. For the majority here, the attempt is made to standardize on "Crewdog".

Many of the words used by the Crewdogs are based on acronyms. Many others are descriptions of what they come to mean, and many are slang terms that have been passed from one generation of SAC personnel to the next, with no one actually knowing why they came to be given those labels.

The following is a good sample, but very incomplete list, of many of the terms that were used everyday by B-52 aircrew members in the Sixties and Seventies. Many of these terms are used throughout these stories.

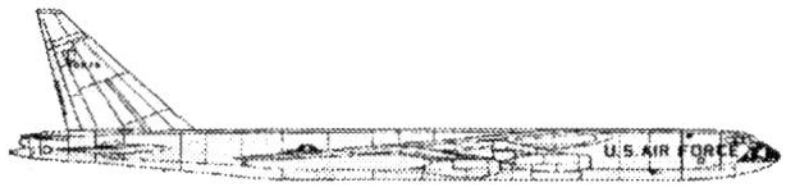

11-DAY WAR – What SAC crewmembers called the North Vietnam bombing operation officially named Linebacker II, because B-52s flew bombing missions for 11 days during that operation.

781 FORM (Air Force Tech Order 781) – A book in which aircraft write ups and corrective actions are kept. Tells number of hours on engines, airframe hours, etc. as well as problems found on the last flight.

AC (Aircraft Commander) – The pilot in charge of the aircraft.

AAA (Anti-aircraft artillery) – Exploding ballistic shells shot at aircraft by ground artillery.

ABORT – To cut short a takeoff, mission, or low level bomb run.

AEW Line – The Airborne Early Warning Line is the point where an aircraft enters the radar coverage of a ground radar system.

AFSC – Air Force Specialty Code – Numeric code for the job you do in the Air Force

ALERT – When crews are kept in a facility on base and are standing by for a quick response to a call to launch in a wartime mode.

ALERT SHACK – The facility where crews live during their week on alert.

ALERT TOUR – A one week period (usually Thursday to Thursday) when a crewmember is stationed on alert with the rest of his crew and other crews.

ALPHA – The radio call sign of the wing commander.

ANDERSEN – Andersen Air Force Base, Guam, a staging base for bombers and tankers

AP – Air Police which later became Security Police – they guarded the base, the planes, and the alert facilities.

API – Amour Piercing Incendiary ammunition that was used in the four 50-caliber tail guns of the B-52D.

AR – Aerial Refueling

ARC LIGHT – A program that deployed SAC B-52 crews TDY to Guam and Thailand to bomb targets in Vietnam.

BAG DRAG – Taking all your flight gear and mission materials from one aircraft to another when the first aircraft has mechanical problems and cannot be launched.

BASE OPS – The place were a crew goes for a final weather briefing and to file flight plans prior to a mission.

BAHT – Thai money. Exchange rate was about five cents equal one baht.

BAHT BUS – Used for transportation from U-Tapao Air Base to Sattahip. Two versions were used: Little Baht busses which were little Toyota type trucks, and the Big Baht bus which was a full size American style bus.

BOMBER OPS – The place where wartime bombing missions are planned by wing staff for the flight crews.

BOMBER PUKES – What tanker crews called bomber crews

BRAVO – Call sign for the wing vice-commander.

BUFF –A nickname for the B-52 meaning “Big Ugly Fat Fellow (Nice Version)”

BUGLE NOTE – A series of grids drawn over a Vietnam map that were used as navigation points for bombing mission planning.

BULLET SHOT – The name of the operation that sent massive amounts of B-52 aircraft and crews to Guam for an intensive bombing campaign of Vietnam.

BURN OFF FUEL – A B-52 aircraft could take off with and be airborne refueled with more fuel than they could safely land with. In emergency situations, the aircraft would have to fly around and use up excess fuel until a safe landing could be attempted.

BUST A CHECKRIDE – To fail an in-flight evaluation given by a member of a standboard crew.

BUY NONE – A wing inspection like an ORI, but without the crews actually flying the mission.

BUY THE FARM – To crash an airplane. Comes from an urban legend that the government had to pay farmers if an aircraft crashed in their fields.

BX – Base exchange. The retail store on an Air Force Base.

C-SQUARE - Nickname for the CCRR event . See definition that follows

CARTRIDGE START – Using black powder filled cartridges in B-52 engines that when fired propelled the turbines enough to spin to start the engine.

CAUTION – A procedure in a flight manual that could result in damage to the equipment if not followed correctly.

CCRR – Combat Crew Rest and Recuperation – The rest period given to Crewdogs when they come off of alert. Usually from the time they are relieved on Thursday morning until the following Monday morning.

CCTS – Combat Crew Training Squadron – The organization that trains crewmembers to fly a particular model of aircraft.

CHAFF – Small pieces of aluminum foil cut to specific lengths to mask aircraft from radars. The length of the pieces is determined by the frequency of the radar. Rope chaff is long strips that would float down to try to knock out high tension wires.

CHANGE OVER – When a new alert crew relieves another crew of alert responsibilities.

CHARLIE – Call sign for the Deputy Command of Operations

CHECKLIST – A lap book of procedures followed to properly perform a crew function on an aircraft.

CHECK RIDE – A flight designed to check the competence of a crew member to properly do his job.

CMF – Combat Mission Folder

CND (COULD NOT DUPLICATE) - Usually written in an aircraft 781 form when a reported problem could not be fixed because the maintenance crew could not find the problem.

CHICKEN SHOT – A famous in-flight lunch for aircrews flying out of Guam which was almost entirely fried chicken and bread.

COMMAND POST – The primary contact point for crew communications, especially airborne aircraft.

COMMANDER'S CALL – A monthly scheduled meeting between the squadron commander and the personnel assigned to a squadron.

CRESTED DOVE – A B-52 that was not airworthy, but was taxied around air bases so that satellites could see more aircraft than were actually capable of flying missions.

CREW CHIEF – The maintenance person in charge of insuring that an aircraft is kept airworthy.

CROME DOME – Early SAC missions where aircraft were kept airborne for 24 hours at a time and were relieved by other aircraft with the same mission before they returned to their base.

CUT FOUR MODIFICATION – A way of cutting four lines on a parachute so that it could be steered by allowing the wind to flow out of the back of the open chute in a controllable manner.

DEBRIEF – After each mission the crew sits and discusses what happened on the mission with each other or with members of the wing staff.

DEFCON – Defense condition. DEFCON = 5 Normal peacetime readiness; DEFCON 4 = Normal, increased intelligence and strengthened security measures; DEFCON 3 = Increase in force readiness above normal readiness; DEFCON 2 Further Increase in force readiness, but less than maximum readiness; DEFCON 1 =Maximum force readiness.

DEMOCRAT – The call sign for one of the HF radio stations that broadcast EMA messages to SAC aircraft in a manner such as "Skyking, Skyking, this is Democrat. Do not answer, break, break…" followed by a series of coded letters.

DOWNTOWN – During the North Vietnam bombing campaign this is what Hanoi was called.

DASH 1 – The tech order for the aircraft which covered all the things a Crewdog needed to know about the aircraft.

DASH 2 (Manual for classified EWO equipment)

DNIF (Duty not involving flying) – A condition imposed by a flight surgeon on a Crewdog that was too sick to fly.

DRAG CHUTE – A large parachute attached to the tail of a B-52 that is released on landing roll to slow the plane down.

DRAIN WATER - When outside temperatures were going to go below freezing the water used to augment takeoff thrust by some B-52 aircraft had to be dumped onto the ramp.

DREAM SHEET (AF Form 90) – A form filled out by officers listing their choices for future assignments and submitted to the personnel office.

E- CREW – Elite crew. A crew that was more skilled than an R-crew.

EAM – Emergency Action Message – A message sent to aircrews to alert them in the time of national emergency.

EAR – Emergency Armed Release – to release all the bombs at one time at the end of a bomb run to insure that they depart the aircraft.

ELEPHANT WALK – Taxing all the alert aircraft down the runway to simulate a launch.

EMERGENCY PROCEDURES TEST – A scheduled or non-scheduled testing of the ability of an aircrew member to pass a closed book test on procedures to be followed in an aircraft emergency.

EWO OFFICER – Electronic Warfare Officer. Later changed to Defensive Systems Officer, the EWO was the crewmember in charge of protecting the aircraft from airborne radar and heat seeking missiles and from land based radar threats such as SAMs and AAA. This was done with a combination of radar jammers, chaff, and flares.

EWO MISSION – Emergency War Order mission refers normally to a nuclear war planned mission

EWOT – Electronic Warfare Officer Training

E&E – Escape and Evasion. Crews were trained on how to avoid capture if they were shot down behind enemy lines.

FALCON CODE – A collection of rude and many vulgar statements that were given code numbers so that they could be said over the radio without having to say the real phrase.

FAMILY VISITATION CENTER – A building that was located outside of the fenced-in Alert area in which families were allowed to come and visit Crewdogs that were on alert

FEET WET – Crossing the coastline and leaving land behind you and water beneath you.

FIIGMO – (Pronounced fig-mo) F*ck It I've Got My Orders, a term used by a short timer who already has orders to a new base.

FLARE – A defensive weapon dropped from the tail of the B-52 aircraft to defend against heat seeking missiles.

FLARE – The maneuver of pulling the nose of the aircraft up to a landing position just before touching down on the runway.

FLY OVER – Flying an aircraft close to the ground over an airfield during events such as an open house or air show.

FORM 90 (Officer Assignment Worksheet – See DREAM SHEET)

FOXTROT – The call sign for the Supervisor of Flying who works with aircraft preparing for launch or recovery from a mission.

FSAGA – First Sortie After Ground Alert. The first time a plane was flown after it had been reconfigured after sitting on alert and was being moved back to a training function.

FUBAR – "Fouled Up" (Nice version) Beyond All Recognition

GEAR LOCKS – Metal pins that were put in the landing gears to insure that they do not retract while on the ground.

GILLIGAN'S ISLAND – A fast food mobile kitchen set up under some palm trees outside of 43rd Base Ops in Guam.

GUAM BOMB – An old beat up and usually rusty car bought because it was cheap.

HABU LIGHT – A small rectangular rechargeable flashlight made by Sanyo that had both red and white lights and was bought overseas and usually carried in a flight suit pocket.

HASTINGS – A stateside bombing practice route

HF – High Frequency radio. Used by crews for long distance communication with SAC command posts

HONEY BUCKET – The non-flushing toilet on a B-52 aircraft, often the object referred to in the phrase "You use it, you empty it."

IP – Instructor Pilot

IP – Initial point, A well-defined point, easily distinguishable visually and/or electronically, used as a starting point for a bomb run to a target

JAMMERS – Black boxes designed to match the frequency of and confuse radar signals that are trying to locate the aircraft

JP-4 – The designator for the fuel used in the B-52D.

KENNY'S STEAK HOUSE – A food establishment outside the back gate of Andersen AFB, Guam

KLAXON – An obnoxiously loud horn, several of which were placed around a base, and were blown to signal an alert response to alert crews

LAJUNTA – Located in Colorado, this was one of the many low level bombing ranges flown by bomber crews.

LAST CHANCE –The activity when the Supervisor of Flying drives around the aircraft to give it a visual inspection before Prior to takeoff on a sortie.

LOW LEVEL BOMB RUN – A bomb run where the aircraft flies low to avoid radar detection

MACV - Military Assistance Command, Vietnam

MADAM T'S – A massage parlor down the road from U-Tapao.

MILK RUN – An easy mission usually flown over South Vietnam with little combat risk.

MISSION BRIEF – Prior to flying a mission each crewmember explains what his duties will be and what he will be doing during a mission.

MISSION PLAN – (v.) The act of planning what is going to be done on a mission and (n.) the finished product of the planning

MITO – Minimum Interval Takeoff or when aircrafts take off one behind the other with very short spacing (usually 15 seconds) between one aircraft and the other.

N-CREW - Non-combat ready crew

NBT – Nav-Bomb Training

NEWLAND – An adult recreation area down the road from U-Tapao.

NOMEX – A flame retardant material from which flight suits are made.

NO NOTICE – Usually an evaluation flight in which the evaluator shows up to fly without anyone knowing that the evaluation would take place.

NOTE – A less important piece of information in a flight manual that must be followed to properly operate the aircraft or equipment, but which will not damage either the equipment or the operator.

O' CLUB – A building restricted to use by officers only where social events take place and meals are served.

OB-xx – "Oil Burner." The 1960s stateside low level training routes, designated by "OB" followed by a number and usually the name of the town where the Radar Bomb Scoring Site (RBS) for that route was located, such as: "OB-1 Wilder" – (Wilder, Idaho).

ORI - Operational Readiness Inspection - A very important Higher Headquarters inspection of a flying unit to measure their ability to perform their mission.

PCTAP – Positive Control Turn Around Point. A point on a war plan which cannot be crossed unless express permission is received from the command authority. Bombers can orbit there awaiting the message until it reaches the point where it must return to base because of low fuel.

PEN GUN FLARE – A small pencil size flare gun in an aircrew survival kit that fires a red flare to signal search and rescue units.

PLF – Parachute Landing Fall, the way Crewdogs were trained to land on the ground to avoid injury when parachuting from an airplane.

POP UP MANEUVER – A maneuver that a low level bomber performs to gain enough altitude to safely release its weapon.

POSTFLIGHT – Performing checklist procedures to safely shut down and make an aircraft safe to sit on the ramp

PREFLIGHT – Checking out an aircraft to see that it is in a flyable condition and that all of its equipment is operable

PRIDE CHECK – The items that were checked when checking out an aircraft oxygen system. It stood for Pressure, Regulator, Indicator, Diluter, Emergency bottle

PUNCH OUT – To eject from an aircraft

R-CREW - A ready crew or a crew that is certified to perform a mission.

R&R (REST AND RECOUPERATION) A period of time given to an aircrew member following or to break up a long deployment.

R&R (REMOVED AND REPLACED) – When a malfunctioning piece of equipment has been removed from an aircraft and replaced with a working one.

RAMP – The concrete taxiway and the parking spots where aircraft were placed

RECALL – An event in which all crew members and essential base personnel are called back to the base.

RECALL ROSTER – A list of personnel who are to be called and the assignments to each as to who else on the list they are to contact.

RED BALL – A truck used to bring last minute equipment repair just before a sortie.

RED DASH – A diagonal line put in the aircraft forms with a red pencil to denote an aircraft problem that needs to be corrected, but does not prevent an aircraft from flying a mission.

RED X - More problems than a Red Dash, an "X" is put in the form to indicate a maintenance 'write-up' code that 'grounds the aircraft' until the write-up is corrected.

RHAW GEAR – Radar Homing and Warning equipment used by the EWO to monitor radar activity directed at the aircraft.

RICHMOND – A stateside practice bombing range

RN – Radar Navigator; once called the bombardier in earlier aircraft, he is the one that is responsible for getting the bomb to hit the target. He uses the bombing radar to know when to release the bombs.

ROACH COACH – A portable food vender truck that drove around the flight line selling snacks and drinks.

RR-112 – A designation for the type of chaff dropped in training from a B-52

S-CREW – A Select crew or Standboard crew.

SAC – Strategic Air Command, the command authority of the entire strategic bomber and missile force.

SAM – Surface to Air missile such as the SA-2 Guidline missiles that were fired at B-52s in Vietnam

SATELLITE ALERT – To pull alert duty at an air force base or station other than the primary base to which crews and aircraft were assigned.

SCM – Start Counter Measures – a point on a bombing mission when the EW starts using his jammers and chaff

SEARCH AND RESCUE – The procedure to find downed aircrews usually in a hostile environment.

SHACK – Dropping a bomb exactly on the target it was intended to hit.

SHOOT A SEXTANT – Use a periscopic sextant to measure celestial bodies to calculate the aircraft's position. The EWO on a B-52 usually performed this duty and the navigator plotted the results on the chart.

SHORT TIMER – Someone who did not have much time left in completing an assignment, such as a TDY tour to Guam.

SIX AND TWO COUNTED AND STOWED - referred to ground locks and bypass keys. The ground locks were the pins that locked the landing gear down, one for each of the main trucks and one for each outrigger--six. The two bypass keys were used to allow front truck steering when they uked the airplane with a towbar.

SIX PACK – The alert truck was a six-seater pickup, long before they became popular as SUVs, and earned this nickname.

SKYKING – The generic call sign of all SAC aircraft when airborne.

SNAFU – Situation Normal, All Fouled Up (Nice version)

SORTIE – A flight by an aircraft.

SPECIAL WEAPONS MANUAL LOCK HANDLE – A handle in the EWO compartment of the B-52D that must be pulled and locked before the RN can drop a nuclear bomb. Part of the check and balance system to prevent one man from having too much ability to do something alone.

STAG BAR – Before the era of being political correct, this was a bar that was off-limits to females including members and spouses of members. It was sometimes the site of female dancers providing entertainment.

STAND UP – A weekly briefing of command staff held by the wing commander.

STANDBOARD – The organization in a wing that was responsible for administering checkrides and emergency procedure tests, among other things.

STATIC DISPLAY – To open an aircraft up for public display for visitors such as at an open house.

T-4 – The electronic warfare officer's ground simulator which duplicated his aircraft position.

T-10 – Bomb/Nav trainer

TACAN – A radio aid to aerial navigation.

TANKER TOADS – What bomber crews called tanker crews

TAT – Target Area Tactics begin when you enter a threat zone.

TAXI BACK – To do a full stop of an aircraft often to let non-essential crewmen off before the pilot continued touch-and-go landing practice.

TCM – Terminate Counter Measures – the point on a bombing mission when the aircraft has left radar coverage and the EWO can terminate jamming and chaff

TDY – Temporary Duty at a distant base.

THE ROCK – Guam or PGUA as it was often called.

TICKETS – Plastic laminated war codes that must be broken open and verified by two officers before the launch of a EWO mission.

TIME HACK – Like in the movies when each crewmember synchronized his watch with others so that everyone's watch shows the same time. We used WWV the National Bureau of Standards as the source for our official time.

TOT – Time over target. The time when a bomb delivery is to be made.

TOUCH AND GO – A practice landing in which the aircraft actually lands on the runway but does not come to a stop but adds power and takes off again.

TTAT – Terminate Target Area Tactics – the point on a mission when the aircraft exits the target (and treat) area.

TWO MAN POLICY – A policy that states that no one person can be in a restricted area alone. There must always be two men there to insure the safety of the area.

TWO OFFICER POLICY – The policy that two officers must verify the codes of a launch order to insure that no one man can launch a weapon alone.

UKE – A tractor-type device hooked up to an aircraft to tow it around on the ground

UHF – Ultra High Frequency Radio

UT – U-Tapao (Utapao) Royal Thai Air Force on the coast of Thailand.

VHF – Very High Frequency Radio

VOR – A radio navigation aid.

WARNING - A procedure in a manual that could result in injury or death to the crew member if not followed correctly.

WATER WAGON – A B-52D that used water augmented thrust to achieve a better takeoff.

WWV – The station that broadcast the correct time that was used as a standard for all service branches.

YOUNG TIGER – The program that deployed KC-135 tankers and crews to Southeast Asia

ZULU TIME – Greenwich Mean Time, or Universal time as it is now called, was a common time system to insure that the same time is used by all operations around the world.

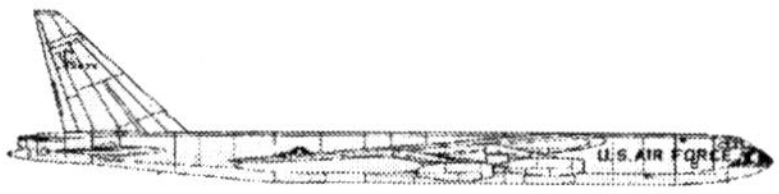

Peace Is Our Profession

It is no small wonder that the life of a BUFF Crewdog is somewhat hard to explain to other people. As members of a B-52 crew, we were assigned to the Strategic Air Command (SAC). Somewhere near the entrance gate of almost every SAC base was a big sign proclaiming to all that enter that "Peace Is Our Profession." Once you entered that gate and went to any operational squadron building on the base, there was normally that other sign hanging in some office that boldly proclaimed "Our mission is to fly and fight, and don't you forget it!" I guess we were expected to fly and fight for peace, and that was what we did. Hence, it was not uncommon to hear the Crewdog's own version of the SAC objective as "Peace Is Our Profession – Killing is just a hobby." Knowing that, it should also come as no surprise that we would also often use the term "SAC Trained Killer" to describe ourselves.

Once checked out in the squadron's model of B-52 aircraft, a Crewdog knew that his association with the BUFF to which he was assigned would be spent doing one of three main missions. One of those missions did not actually involve flying, but it still required daily contact with the aircraft. That mission was protecting our country by being a part of the Nuclear Alert Force.

The other two missions actually involved flying the BUFF. The most understood and most publicized flying mission was the flying of honest-to-goodness combat missions. In the late Sixties and early Seventies, TDY (temporary duty) crews were sent from home bases in the states to fly those missions in Southeast Asia. The first deployments of this kind were called "Arc Light." Later on, a massive buildup of B-52 aircraft in Guam and Thailand was initialized under the name

"Bullet Shot." Combat missions were normally 12-plus hours in duration when flown out of Andersen AFB on Guam. A combat mission launched from U-Tapao Royal Thai Airbase in Thailand was normally about a 3½-hour mission. Almost all of these missions were designed to bomb targets located in Vietnam. Those were the flights that most people read about and see on television. They are not hard to understand.

The most unpublicized and least documented flights were the stateside training missions. Those were the ones that appeared to be a chore most of the time. Even when the conventional (non-nuclear) bombing was going on in Southeast Asia, most of the training missions flown in the states were designed to train aircrews to carry out the nuclear mission of SAC, the SIOP (Single Integrated Operations Plan). These training missions were flown in a high-low profile, which meant that BUFFs would fly high altitude until they got close to the target area, and then they would drop to a low altitude profile to escape enemy radar detection from enemy air defense systems.

The bombing missions in Southeast Asia required flight procedures and skills that were totally different than the skills required for these nuclear mission profiles. The bombing in Vietnam was done from high altitude, with the missions usually flown between 35,000 and 40,000 feet. Many of these missions were controlled by ground assisted radar sites that directed the B-52s to the target area and instructed the bombers when to drop their bombs. Unlike the World War II movies, it was rare when a bombardier actually looked through a bombsight to line up the crosshairs on a target he could see on the ground. When not directed from the ground, radar imaging was the primary method of selecting the target.

Included in this collection are stories about all of these three types of duties, along with first-hand reflections on many other aspects of the day-to-day life of the Crewdogs.

EWO Training School Patch

Becoming a Crewdog

Vince Osborne

How does a guy get started down the road to Crewdog? Here's how I did it. In November of 1965, I was assigned to pilot training in Enid, Oklahoma. I had graduated from Colorado State College in Greeley Colorado and had received my commission in the Air Force as a Second Lieutenant through the Reserve Officer Training Corps (ROTC). I was ill prepared for the intensity of flying for the Air Force. I was too timid. I lacked confidence. Although I did some of the harder things well such as landing the T-37 twin jet trainer and recovering it from spins, there was a great deal of airmanship that was beyond my grasp. Shortly after I soloed the T-37 "Tweet" a couple of times a Flying Evaluation Board (FEB) recommended that I be "washed out" of pilot training. At the time about 25 percent of those entering pilot training were being washed out. Given enough time, I could have flown safely. No amount of time would have given me the personality that the pilots were looking for.

Most of us who washed out of pilot training were given the opportunity to go navigator training school at Mather AFB (affectionately referred to as Mother AFB) near Sacramento, California. I learned there that many of the attendees had also been to pilot training and had washed out and laughingly called pilot training

"Navigator Prep School." The shock of being "WASHED OUT" made me vow to never fail at a school again. I knew I was no Einstein, but I knew I knew enough to make it through anything if I really tried. And so I hit it. There was a lot to learn. We had to be able to identify all the primary navigational stars (52 of them I think) and would sit in a small planetarium on base and learn little catchy phrases to help us identify them. We learned constellation and individual star names as well. We had to learn to take and send Morse Code. Detailed weather knowledge such as identifying cloud types was required, and the weather test was believed to be the hardest test of the school. The staff had a predictive measurement of some type that said I would fail the test miserably. I decided that there was no way I'd let it happen. I studied very hard. I got an 88 on it. That was not a chart buster, but a comfortably passing grade.

The flying was also tough. We did our training in a twin-propeller-engined trainer designated the T-29 "Flying Classroom." The T-29 was accepted by the Air Force in 1950 and was slow compared to most of the military jets that graduates would transition into. It was built very similar to a commercial plane called the Convair 580 and the USAF C-131 transport aircraft and had individual stations and instruments for each of the 10 navigator training students who could take turns navigating the aircraft. One has to be extremely well organized to take a fix, (determine the aircraft position of the aircraft over a point on the earth), put out a dead reckoning (DR) position, and prepare to take the next fix to adjust the DR position to a real fix. There was a demanding physical training element to the curriculum. It included "Combative Measures." I liked it. I did well in it. It combined karate, judo and other martial arts that might be useful if close to an enemy soldier.

I didn't finish at the top of my class, but I did complete the training successfully, and won my wings. Later, a dear little boy would ask me, "Are those your silver wings?" They were. And I didn't want to lose them, ever. There were three routes out of navigator school - Operational assignments, NBT, and EWOT. Operational meant going directly to a flying squadron as a navigator. These assignments were normally with the Military Airlift Command (MAC), which was just growing out of its roots as the Military Air Transport Service (MATS). The Navigation Bombardier Training (NBT) was also located at Mather and was preparation to work in the downstairs navigation compartment of the B-52 heavy bomber aircraft. The third choice was EWOT stood for Electronic Warfare Officer Training. What was taught there was little understood by most of us. We knew it was a very difficult school from an academic point of view, and we knew that if we failed at it, we would face a flying evaluation board that would end our flying careers before they really got started.

Each Nav class was given a list of available assignments, based upon the current need of the Air Force. The accepted procedure was to allow graduates to select their assignments from the list based upon their ranking in their class. I elected to go straight to an operational job. I wanted to fly in the Douglass C-133 Cargomaster. The C-133 might appear at first glance to be much like a C-130 Hercules, but it was larger. It turned out to have a poor safety record, and didn't have the staying power that the C-130 has proven to have. However, "they" decided I would go to EWOT. I don't know how the decision was made. I went to a Captain or Major whom I thought had something to do with the decision and pleaded my case, but he told me I could do it, and I resigned myself to it.

Becoming a Crow

The official mascot of the Electronic Warfare profession is a black crow standing with bolts of lighting held beneath one wing and another set of lighting bolts shooting from his upheld claws. During World War II, the Allied Radar Countermeasure operators used the code name "Ravens" and employed receivers and transmitters to monitor and jam threat frequencies. Military jargon later changed "Ravens" to "Crows". In the Air Force, these operators were given the official name of Electronic Warfare Officers (EWO). Although the path to become an EWO required completion of Navigator Training School, the aircrew mission of that position was vastly different and had little to do with navigation.

Inside the chain link fence with three rows of barbed wire on top was a guarded entrance to the school. Of all the Air Training Command facilities on Mather AFB, this was the most secure. Even though it was definitely a school, the atmosphere seemed more related to the real world than had the previous schools. And yes, it was hard. There was a lot of basic circuit design study. We had to learn color codes for resistors and how the newly developed electronic marvel called the transistor worked. Some of the guys in the class were electronic engineers and had little trouble with that portion of the curriculum. I had had only one basic electricity course, and I found it very demanding. I stayed in the building late almost every day to study. I studied at home, I got together with other guys who volunteered to help me, and I talked with the teaching staff to get extra help. I wasn't going to wash out of this school either.

Besides book work and the flying, there were two different kinds of flight simulators we had to master. One was a bomber defense simulator called the T-4. It simulated quite accurately the Electronic Warfare Officer's compartment of one of the models of the B-52. The other simulator was the T-3 and it simulated a generic electronic intercept aircraft or a snooper or a reconnaissance aircraft. I made it through the bomber defense T-4 quite well. I needed some extra missions in the T-3 to get through that. I also got through the flying missions we flew in an airborne training aircraft equipped to do real airborne signal intercept. The aircraft flown was a modified T-29, called the ET-29. Instead of navigation instruments, this flying classroom had positions containing signal intercept and radar jamming equipment. With the flying, simulators and classroom training successfully completed, I was finally declared an electronic warfare officer—and was darned proud of it.

The B-52D

The "Dog" model. At one time the phonetic alphabet used by the Air Force to make letters of the alphabet clear over the radio started as follows: Able Baker Charlie Dog. When talking on a radio (especially HF) and referring to anything that had letters in it or when asked to spell out a misunderstood word, the phonetic alphabet was used. Later, the International Civil Aviation Organization (ICAO), which made English the worldwide aviation language, created its own phonetic alphabet using words and agreed upon by nations speaking a variety of languages. Each word is supposedly unmistakable. It now begins Alpha Bravo Charlie Delta. But, at the time it made more sense to refer affectionately to the D model as the Dog model. As in "Oh, you'll be flyin' dogs?"

In 1968, when I was assigned to the B-52Ds at March Air Force Base in Riverside, California, I had not yet seen that model of B-52. The reason was because they were all mostly deployed overseas in the Southeast Asia Theater. From Mather, I went down the state to Castle

AFB in Merced, California, for Combat Crew Training School (CCTS). There we flew B-52E and F models. The B-52Ds, however, were configured much differently from the E and F models. Although they were older, the D models had been greatly modified to perform the "conventional" (a euphemism for non-nuclear) role in Southeast Asia, and had the latest electronic warfare equipment. It would be my job as a new electronic warfare officer to operate that equipment and protect the aircraft from enemy threats such as Anti-Aircraft Artillery (AAA) and Surface-To-Air missiles (SAM), and enemy fighters.

As a part of the pipeline of training leading to the operational flying of the D, I also went to Survival School at Fairchild AFB in Washington state. Besides being the site of Survival School, Fairchild was also the home of an active duty SAC unit that flew B-52s. There they had the Ds. I first saw them sitting on alert at night. They had new polyurethane black paint on the under sides, and camouflage on top with the required fuselage markings in a deep red. They looked like genuine war machines, and I was in love. I couldn't wait to get into my own Dog model and learn the skills needed to protect that aircraft and its crew and mission from the electronically aimed and guided threats of the Viet Nam Theater of operations. My chance would come in 1968.

Crew R-04 in our version of crew "Party Suits" tailored in Sattahip. While most crews preferred traditional flight suit design, we fashioned ours after "Startrek" uniforms. Back Row (L-R) -Tom Adcox, Dave Kerr, Denny Scruggs. Front Row (L-R) - Tommy Towery, Doug Bunch, Rich Hayes.

My First Crew

Tommy Towery

Once you are officially checked out in the aircraft, you are finally ready to join a real crew. There's something special about a first crew in the same way that there is something special about your high school graduation class. Those people seem to always have a place in your memories. Before becoming a part of a crew, we all went through training with various other people who were also learning how to master their aircraft positions, but were never a part of a real crew. In initial training, almost each mission was planned and flown with different students, and you never seemed to identify with them as a whole. We never had to share the day-to-day challenges with them the way that we are forced to share our lives when we were assigned and became an integral part of a real crew.

My first crew was Crew R-04. In crew terms, "R" crews were "Ready Crews". We were all mostly recent graduates from CCTS and were thrown together as a crew created in a rush to be deployed to Southeast Asia. This was shortly after the Vietnam War had ended, and even though all the fighting and bombing was over, the political situations and tensions were still uneasy there; therefore, we were required to maintain a bomber presence.

My newly created crew was formed at Carswell AFB, Texas. We were from the 20th Bomb Squadron of the 7th Bomb Wing stationed in Fort Worth, Texas. The 20th BS had the nickname of "Buccaneers" and is most famous for one of its past commanders – Col. Billy Mitchell. Most of the crew were also new to the B-52 aircraft for which we were now qualified to man.

It seemed like our crew was just barely assembled before we were deployed to Southeast Asia – or more specifically, Guam and Thailand. Due to an accident at home, something to do with a broomstick and a poke in the eye I seem to remember, we deployed without our assigned Radar Navigator. It would be a few weeks later when he would join up with us, after we had arrived at U-Tapao Royal Thai Airbase in Thailand.

Just as you remember your first girlfriend's name, it is easy for me to remember the names of those that made up my first crew. It consisted of Denny Scruggs, Aircraft Commander (Pilot); Tom Adcox; Copilot; Dave Kerr, Radar Navigator (Bombardier); Rich Hayes, Navigator; myself as the Electronic Warfare Officer; and Doug Bunch, our tail gunner (officially called the Fire Control Operator). Our aircraft was a B-52 D-Model, which meant that our tail gunner actually sat in the tail of the aircraft and manned four 50-caliber machine guns that fired 1250 rounds of Armor Piercing Incendiary (API) ammo. Later models of the B-52 had the gunner sitting up in the front crew compartment and allowed him to use a radar and television optical system to see approaching fighters.

I can't even remember the six of us flying that much together as a crew stateside before we were deployed, but we must have done that some. Training missions had a tendency to be crowded and confusing when you add in all the instructors and Stand board (flight evaluator) crews for check flights. I guess everything before deployment was kind of pre-climatic (if that is a state) to the thrill of actually going to an overseas base.

Most of us had missed going on an Arc Light deployment and were excited to get to make the trip. I personally had been grounded from flying duty with kidney stones before I ever got checked out in the D-Model and had been assigned to the wing staff – first in the Combat Intelligence office and then the Penetration Aids (Defensive Systems) office. At the time that I was medically grounded, I had just pinned on my Captain's bars, and because of the need for Crewdogs that could fly, I was given the office job, and the Lt Col that worked there went back on a crew. A few months later I was deployed to Guam when Operation Bullet Shot kicked off. While at Guam, I started working on the paperwork to get a waiver to return to flying. During my second six-month TDY to Guam, I got word that my waiver was approved, and I was able to fly two missions as a staff officer before Linebacker II kept most of the staff from flying as extras.

Anyway, it was after my return from that second ground staff trip to Guam when I finally got the medical waiver and was allowed to fly again. I got back into training, and since it had been almost two years since my initial B-52 training at Castle AFB, CA, I was one of the first members of a CCTS (Combat Crew Training School) class for D-Models in the flight school that had been moved to Carswell. I went through the full B-52 initial training program again, and when I got checked out, I was assigned to Crew R-04.

I think I was the only one of the crew that had spent a lot of time in Guam and Thailand, but as a crew, we all looked forward to what the places had to offer. Unlike my Bullet Shot tour, this one was only scheduled for three months, instead of six.

On our deployment to Guam, we missed the entire day of July 4th. We took off on the 3rd from Carswell, and landed in Guam on the 5th. We crossed the International Date Line, and so for all practical purposes, we never saw the 4th.

Denny, our pilot, was given the nickname "Scruggo" because of his last name. We made up a sign that we hung on his door in Guam that read "Scruggo's Bar and Grill," and that became the running joke for the entire crew. He was new to the B-52, but he was a combat veteran, having flow O-1 Birdogs during a tour in Vietnam. He had also flown HU-16 Albatross seaplanes in Newfoundland before being assigned to Texas. Our other running joke with Denny was to see how many times we could get him to have his hair cut. He was a stickler for military standards, and the rest of the crew would get together as a group and start talking about how long his hair was getting. He'd act like he wasn't paying any attention to us, but the next thing you know, he'd disappear from the Alert Shack and come back with a new haircut.

Dave was affectionately known as "Mad Dog" Kerr. I don't know who put that on him or when, but he was stuck with that one. Perhaps it was from the similarity of the pronunciation of his last name to "Cur" which by dictionary definition is "a mixed-breed dog, especially an ill-natured or worthless one". Dave was the most mature of the group and the highest ranking, but he was known to have a good time at other's expense after he had downed a few drinks.

Tom was from Duncan, Oklahoma and was a true "Okie". His farmhand strength proved invaluable during one particular low-level flight when the aircraft autopilot kicked off and the plane experienced a "hard-over" maneuver, which pointed the noise of the aircraft toward the ground. We were at 200 feet altitude on a low-level bombing run when that happened, and Tom just gripped the yoke like Big Bad John and pulled up and manhandled the big bomber back into a straight and level flight.

Richard Hays, the navigator, was never given an Air Force nickname, but was just called by his civilian nickname "Rich". He was one of those quiet people that seem to meld into a crowd, but beneath that disguise was another persona. Rich was the kind of guy that would pull a joke on someone and you'd never know who did it. He was the last one that would be suspected, so he could get away with murder – the kind of guy that could s**t in your box lunch and not even laugh out loud when you opened it, but inside would be rolling with hysteria.

I was the EWO. I was "Tango". Actually, one of my friends at Carswell started calling me "Tango-Tango" which came from the phonetic alphabet spelling of my initials "TT" and was later shortened to just "Tango." I am not sure how the others on my crew interpreted me. Personally, I feel that I was one that was excited about my job and the challenges of the Air Force. I had my shares of ups and downs during my career and sometimes felt that I did not get a fair shake on some things. Maybe we all did in one way or another. I had washed out of pilot flight training, along with 50% of my class, and I had found that I enjoyed the electronic warfare officer roll. My frequent attacks of kidney stones twice threatened to cut short my planned Air Force flying career and caused me to get assigned to some ground jobs that I never felt comfortable with. I was really excited about finally getting back onto flying status and assigned to a crew. Since I had already been to Guam and Thailand in staff jobs, I became the guide for the rest of the crew when we were sent there.

Doug, our gunner, was young. That's the biggest impression I have of him. He was one of the first that went straight to a B-52 assignment after leaving Fire Control Operator training. He did his job, he fit in great with the rest of the crew, and we all respected him and the job he did in spite of his youth. One memory of him was a story he told about returning to his hometown in North Texas after our deployment to Guam. He was talking to some of his high school buddies and one of them was bragging about how he had gone to Dallas for the weekend over the summer and how exciting it was to go somewhere big after growing up in a small town. When Doug was asked what he did for the summer, he said he had spent it in Guam and Thailand. I don't think his buddies' trip to Dallas sounded so exciting, then.

That was the makeup of the original crew, and we enjoyed flying together. We went from an "R" crew to an "E" (senior) crew quickly and did a good job. But, like families of brothers separate as they grow older, so did the crew as we all became more proficient in our duties. Copilots become pilots, navigators become radar-navigators, some of us become instructors, and then some continue on to standboard duties. It doesn't all happen at once. It's insidious. Slowly, one-by-one, the positions manned by the people I came to know as crew R-04 changed. One person left, and another took his place. What were originally six become five and a new guy. Then before you know it, there are only four of the original ones, then three. Before long, the crew number is the same, but none of the original members are there.

The odd thing is that today when I look back on the other crews to which I was assigned, I remember the people and their positions, but I can't recall all the crew numbers that I have had. Even today, if anyone were to ask me what my crew number was when I was in the Air Force, I would immediately spit out "R-04."

The Aircraft Bible - The Dash-One

Tommy Towery

One of my earliest memories of my first B-52 assignment was being issued my first flight manual. The manual was held together in a corner with the biggest industrial strength staple I had ever seen and it took a pair of pliers and a screwdriver to remove the staple. The front page had a picture of the aircraft and printed in the top right corner of the page was "T.O. 1-B-52C-1." The "T.O." stood for "Tech Order" but the manual was affectionately called the Dash-1 and was the Bible of how the things in a B-52 worked. Periodically updates would be published to either clarify something or to cover new rules or equipment changes. When I was given my first one in 1970, I was issued a basic manual dated 20 June 1967 that covered both B-52C and B-52D aircraft. I was also given 21 packages of changes to the basic manual that had been issued since that first printing. As of Change 34, dated 15 April 1974, the total number of pages in the Dash-1 was 1093 pages, over a ream of paper printed front and back.

I remember taking this monstrously large stack of pages home and spreading it out on my living room floor. The problem with making all the changes is that you could not just add the last change; you had to do all of the changes starting with the first one. You had to do them in order, because no change was a complete new manual by itself. On the front page of each new stack of changes were the words "Change Notice. Latest changed pages supersede the same pages of previous date. Insert changed pages into basic publication. Destroy superseded pages." For most of us, to throw the pages into a trash can passed the test for "destroy."

Many of the changes were pages that you removed pages from the basic manual and replaced them with new ones that were printed for the change. Many changes were "pen and ink" changes that instructed you to go to a certain page and draw a line through the old paragraph and write in the new change by hand. I also remember that one time a mandate came down that "Effectively immediately all pen and ink changes to the flight manual will be made in pencil." That always struck me as being funny.

As I progressed through this stack of papers, I would find myself writing a long pen and ink entry on a page for one change, only to turn around five minutes later and throw that page away when I replaced it with a brand new page printed in the next change.

The Dash-1 was kept in a large three-ring notebook binder and the basic manual's (before changes) pages all lined up nice and straight with all the edges even. It seemed that with every change that was printed, the holes were punched in a different location than the change before. Many times the holes punched in the paper would not line up with the clips of the notebook at all. Sometimes they were punched for nine-ring binders, sometimes, six, and sometimes three. At the end of the evenings work on constructing my first Dash-1 and bringing it up to date, the bulging notebook's rings had been unsnapped and re-snapped so many time that they would hardly stay closed, and the edges of the papers inside the notebook looked like a scrapbook made of different sized pages.

Strategically placed throughout the manual were three types of things that demanded special attention for the safe operation of the aircraft. These were warnings, cautions, and notes. Warnings were operating procedures, techniques, etc., which will result in personal injury or loss of life if not carefully followed. Cautions were operating procedures, techniques, etc. which will result in damage to equipment if not carefully followed. Finally, Notes were an operating procedure, technique, etc., which is considered essential to emphasize.

The Dash-1 was not just a book to help you fly in the B-52; it was also an object of inspection for every checkride Crewdogs had to take. During those checkrides the evaluator would take out the Dash-1 of the person being evaluated and insure that all the latest changes had been made, and that all the pen and ink notations were done properly. It was a nightmare when a new change came out on the day before a scheduled checkride, because you had to have it entered before you crawled into the aircraft.

It was a combination of this fear of not having made all the changes properly and of the poor holding power of the massive three-ring binder that gave me the inspiration for one of my non-documented procedures. Instead of placing the Dash-1 into my flight briefcase spine first as most people did, I put it in open end first. The tension on the clips of my rings was so bad, that when the evaluator would pull it out of the briefcase, the rings would pop open and all the pages would fall loose into the case. Once that happened, the evaluator usually looked around to see if anyone had seen him dump all the pages, and when he felt comfortable, he would shove the Dash-1 back into the briefcase and never check any of the changes. This got me through a lot of those inspections.

During each annual evaluation you were also required to take a test on your knowledge of items in the Dash-1. Actually you had to take two tests. The first one was a closed book test that covered emergency procedures that you were required to commit to memory. Even after over 30 years since I last set foot in a B-52, I can remember things like “Arming handle rotate, trigger squeeze” for firing the ejection seat and “Bailout and pull parachute arming knob” for the things to do with a manual bailout. The number of rote memory emergency procedures you had to memorize was dependant upon your crew position. A pilot, for example, had to have memorized things that would allow him to do the correct steps in case of emergencies dealing with flying the aircraft. Since other crewmembers could not control those things, they did not have to memorize the same ones.

The other part of the test was an “open book” test that tested the person being evaluated on how easily they could find out information about their crew compartment and equipment and other such things by knowing where to look in the Dash-1. This was normally a 100 question exam, and heaven help the poor soul that did not get a passing grade on this test.

Certain specific materials were extracted from the big Dash-1 and published in a smaller manual, the inflight checklist. The checklist was inserted into plastic pages and had both yellow and white sheets of paper and was made small enough to easily be carried around. In it were the normal steps needed to be taken to do the crew position job. The checklist was divided into sections such as Before Exterior Inspection, Interior Inspection, Starting Engines, Taxi, and other items such as that. Each flight was performed by the checklist, and failure to properly follow it could result in failure of a checkride or worse - personal injury or loss of life.

Like many other things that I now wish I had kept, my B-52 Dash-1 went away during one of the many military moves. I somehow want to remember that at the time I left Carswell AFB, Texas, there was a shortage of flight manuals and so I had to turn it in for a new Crewdog to get. It didn't seem fair that he would not have to go through the pain and labor that I had to when I first arrived on station and had to sit in his own living room floor and post all the changes to the basic manual.

Peacetime

"SAC Warriors"

Bomb-Comp Patch, Blue Ribbon Crew Patch,
and Mathis Trophy for High Altitude Bombing Competition.

The B-52 Peacetime Missions

Charles N. Brown

(The below article is extracted from "The B-52 and Linebacker II" originally printed in the Air Force Museum newsletter and reprinted here with special permission obtained from that publication.)

For many years in SAC, the typical mission routine at US B-52 bases was something like this: The average unit supported about fifteen B-52 aircraft. About a third of these, with their crews, were on ground alert 24 hours a day. Crews were on alert at a special area of the base where they lived, ate, slept, and had their aircraft parked. They operated there for a week at a time. Aircraft were fueled, loaded with nuclear weapons, and systems were checked and positioned every day for speedy response. The normal alert time requirement was to be airborne within fifteen minutes if an alert sounded. When national defense readiness conditions increased, this response time dropped to just five minutes. Individual B-52 aircraft sat alert thirty days at a time. They were then put back on the flight line for regular flying. Ground alert ended in September 1991.

A typical training flight in a B-52 was six to ten hours long. It could have been day or night. It included all the elements of what would be encountered on a real mission: navigation practice, aerial refueling practice, low level flying and low level bombing. This bombing was scored electronically by fixed or mobile bomb scoring sites, since no weapons were carried on training flights. Occasionally high-altitude bombing was also accomplished, and at night, celestial navigation may have been practiced. The end of the training flights usually included a few practice instrument approaches and "touch-and-go" landings for the two pilots.

From 1958 to 1968 a portion of the B-52s were loaded and kept on airborne alert at various times to guarantee survivability of some aircraft in case of an undetected nuclear attack on the US. The typical "Chromedome" route was up the eastern edge of Canada, past Newfoundland and Iceland to Thule, Greenland, up in the Arctic Circle, then across to Alaska and back down to the western continental US. In addition, one B-52 was kept in orbit over Thule Air Force Base and radar site 24 hours a day. The author served at Thule "Iceman" control in 1962 and 1963. This airborne alert mission placed heavy demands on crews and the aircraft. Maintenance and support services were large. These airborne alerts were stopped in 1968. Probably a significant factor in this decision was an accident involving a B-52. An interior fire started from a hole in a hot air duct on the aircraft orbiting Thule. The crew tried to land at Thule Air Base but crashed short of the runway. It went through the ice in the bay and sank. The US recovered the nuclear weapons, but Denmark, which controls Greenland, was concerned. The international political ramifications of a potential nuclear accident were brought to a head, and so flights of nuclear loaded bombers were suspended.

Discussion of the B-52 would be incomplete without touching on one stable mate, a plane that did a lot of work without much glory. This is the Boeing KC-135 "Stratotanker" that replaced the propeller driven KC-97. The KC- 135 is a gas station in the sky and was an integral part of not only the B-52 mission but also the entire Air Force global response and deployment mission. This aircraft was designed for high speed and high altitude refueling. It is equipped with a telescoping boom. This workhorse of an aircraft was reskinned in the 1970s and re-engined in the 1980s. It will probably be around as long as the B-52, and maybe beyond, because dozens of other aircraft these days, especially tactical fighters, are refueled from tankers. The KC-10 is a newer tanker addition.

Most SAC B-52 training flights had a scheduled practice refueling with a KC-135. Refueling rendezvous times and places were prescheduled. The aircraft come in to the initial point from opposite directions, 1,000 ft apart vertically and offset a couple of miles. The radar operator on the bomber called the turn for the KC-135 which then made a 180 degree turn to roll out a couple of miles ahead of the B-52. The B-52 gradually closed the gap and climbed up to the tanker's altitude. Tucking that big B-52 up under the fuselage of the KC-135 so the boom could be attached was one tough, sweat producing maneuver. Pilots, even experienced ones, who were new to the B-52 refueling, thought they might never master it. By ten flights, however, they usually qualified by staying hooked on to the boom for five minutes without a disconnect. Soon, a pilot was expected to hold on for fifteen minutes, which is the time required to transfer a typical combat load of fuel. That produced more sweat. Lights on the belly of the KC-135 gave positioning signals to the B-52 pilot. The acceptable envelope for this close formation position was only a few feet in any direction. Out of those limits the boom automatically disconnected and retracted to prevent damage to either airplane.

The B-52 is not what one would call a pilot's airplane. It's not flown for fun because it's basically a big military weapons platform crewed by trained professionals. Joy comes in doing the best possible job as part of a team. One pilot described the B-52 rather lightly in these words: "The aircraft has enough aluminum and steel to make 20,000 garbage cans. If all the wire and cable in it were laid end to end, it would stretch a hundred thousand miles. The combined power of the engines is equal to 12,000 locomotives. And it flies like 12,000 locomotives pulling 20,000 garbage cans on the ends of 100,000 miles of wire." That's an exaggeration. It may not be easy to fly, but it's not that awkward. In fact, if you take away the bomb loads and reduce the huge amount of fuel on board, this B-52 will really perform. Here is a 164,000-185,000 lb empty weight machine that's normally weighted with nearly twice that in fuel and weapons or another 300,000 lbs up to a total of 488,000 lbs.

Dropping the Big One

Walter J. Boyne

I am not a high-time B-52 pilot, but was checked out as an Aircraft Commander with the 4925th Test Group (Nuclear) at Kirtland Air Force Base. The 4925th was manned for the most part by an elite band of veteran instructor pilots from the B-47 training program at McConnell Air Force base, who had since accumulated several thousand hours in the B-52. I don't think any of them would take offense if I noted that they were also a somewhat wild and crazy bunch, known to take a snifter of Jim Beam on occasion.

In 1962, the United States and the Soviet Union were close to signing an accord that would ban nuclear testing in the atmosphere. Both nations launched quick programs to do some final air-drops to test fusing and firing of existing weapons, and to experiment with new systems for future weapons. Thus it was that the 4925th was suddenly called upon to deploy to the Pacific to engage in Operation Dominic, a highly complex program for live testing of nuclear weapons delivered by both ballistic missiles and by air drops.

I was too junior to be a regular crew member, but acted as assistant operations officer and got to fly on mission in which we dropped a 4.5 megaton weapon, an experience no one could ever forget.

We took off from Barber's Point, Hawaii, heading southeast to an area outside of Christmas Island, where a host of ground stations, surface vessels and orbiting aircraft awaited us. The aircraft were like gears in a clock, all moving around, all intended to be at a precise spot at the moment the huge weapon we carried detonated.

The procedure was fairly simple. The night was absolutely pitch dark from horizon to horizon as we entered the test range and were cleared for to drop. Our radar navigator was a World War II veteran, Major Blake White, no intellectual at all, and as profane and rambunctious man as you would ever meet. He was, however, a master of the radar set, and a brilliant bombardier.

The drop technique was simple—no breakaway maneuver was necessary because the bomb was parachute retarded. We did not even use the anti-glare shades that were provided or put on sunglasses.

Blake called "Bombs Away" then said "I'm coming up to watch." He put his set on stand-by and came up the short ladder from his station to stand behind the pilot.

About forty-five seconds after "Bombs Away" the Mach meter gave a little shudder as the shock wave hit the aircraft. Then suddenly, the entire black world outside turned into brilliant daylight, brighter than the sun could do, when the bomb went off. It was incredible. What had been pitch black night was now intensely bright day, from horizon to horizon. And it was not a flash—the light persisted and persisted for what seemed minutes. It was an almost bewildering experience—and then Major Blake White came up with a remark that I'll never forget. He said, quietly "Every head of state should be forced to see this."

And he was so correct. Heads of states, and generals and admirals and congressmen and everyone in a position of power needed to see exactly what a nuclear explosion was like. In 1962, probably no head of state had ever witnessed a nuclear explosion on anything larger than a seventeen inch television set. There was no way to convey the awful power of an atomic weapon without seeing it. To this day, I remain convinced that it would be worth what ever harm it might do to the atmosphere to bring all the major leaders of the world to a place where they could witness a live nuclear explosion, and make them realize both the power and the danger of nuclear weapons.

Capt. Denny Scruggs at the B-52 pilot's position.

Say Cheese, Please

Denny Scruggs

During the waning days of post Vietnam and entering the late 70's, the Strategic Air Command pursued new and innovative roles to justify keeping the old Dog model B-52 on the ramp and in the skies. One of these experiments was the concept of using them for sea surveillance missions. This mission entailed flying an overwater grid and searching thousands of square miles of ocean looking for enemy vessels. Once identified, the B-52 would make low level passes, called a "RIG" maneuver, and take hand held photos of the vessel to bring back to the base for photo intelligence. That description is very simplified but basically outlines the process.

At the heart of the process is the taking of the pictures. Crew 04 was selected in the early trials to conduct a "proof of concept" sortie. My copilot, a very smart and competent math major from Oklahoma State, had very little experience with handheld 35mm photography, but the copilot was the one designated to be the photographer for these missions because the windows in front of and beside him gave him the best viewing and shooting angle. He picked up the camera gear after mission planning and we preflighted the Buff and got on our way out to the Bermuda Triangle. (It should have told me something when the target was in the BT).

After flying the grid a while we looked down and found our target. I dropped the plane down just above the salt spray off the ocean waves and positioned it to take our pictures. Copilot Tom was like a fashion photographer while snapping away and winding the film. Making comments like "Oh Baby, that's good. Oh sugar, looking fine. Oh man, gonna like that one" he continued until the camera's film counter said 36 shots had been taken and then he stopped clicking and carefully put the camera back in the bag.

My heart sang as we climbed the Buff back to altitude. We would be praised. Our names would be recorded in some AF history book. Spot promotions perhaps would come back to the ranks of the Air Force. Crew 04 had single-handedly saved the B-52D model from obsolescence and given new missions and new life for an old airplane, while opening the door to let enormous amounts of money for the program be approved by Congress. Life was good.

Lots of high rollers met the plane after we landed, including good old Colonel "You're only as good as your Last Mistake". We were all escorted to a debriefing area complete with an audience which included a General officer or two or three - my mind blanks this out now.

I remember hearing someone ask the copilot for the camera. I remember seeing the copilot handing the camera to the photo intelligence debriefer. "Now to get the film out," he said to no one in particular. I remember the sound the film door made as it popped open. I remember the silence that filled the room as the photo intel officer peered inside the camera. I remember some general asking what is wrong and "Where is the film?"

Somewhere the birds were singing, but at that moment at Carswell AFB, the sun stopped shining. The clouds gathered, thunder rolled, and life was not good any more. For someone, the mighty "someone," had forgotten to put film into the camera.

Let Sleeping Crewdogs Lie

Wayne Shimet

My first Crewdog assignment as an Electronic Warfare Officer was flying on B-52G models out of the 51st Bomb Squadron, located at Seymour Johnson AFB, NC. During the spring of 1976 our crew got the opportunity to participate in a joint exercise with the US Navy.

Most peacetime B-52 missions were around eight hours in duration. This one was a 16 hour scheduled flight and since we took off at midnight, it was a killer. Our task on the exercise was to find a US aircraft carrier task force that was cruising somewhere in the South Atlantic. Reconnaissance was not the normal mission of SAC bombers, which were designed to fly to and deliver bombs on a specific target that was pre-assigned prior to takeoff. We set up a grid of the area so that we could systematically search the thousands of square miles of open ocean trying to find the carrier task force and not just fly aimlessly over open water.

As the mission progressed we spent hours trying to find the carrier task force and came up with nothing for our efforts. The Navy had ordered suppressing all communications and electronic emissions within the task force in an effort to make our task a bit sportier. We followed our grid plan, but could not locate the task force either visually or by radar. All of a sudden my warning receivers at my EWO position lit up, indicating that there was a fighter was at our 8 o'clock position. I notified the crew of the radar contact and we adjusted our heading to aim directly for the aircraft that was transmitting the signal. While flying toward the signal, the radar navigator soon picked up the task force on his radar and we successfully found and over flew the ships.

It seems that an electronic technician was making some repairs on the radar of one of the carrier's aircraft. When he finished his repairs, he forgot about not being allowed to turn on any emitters and simply powered up the radar on the aircraft carrier deck to check out his work. As soon as he turned the radar on, we had their location.

We made a few low passes over the carrier task force, rocking our wings and we all had a good laugh for having caught the Navy and even in the way that we found them. Our euphoria was short and we were soon overcome by exhaustion as we settled in for our long flight back to our home base. One by one, the entire crew proceeded to fall asleep. We were at altitude and on autopilot so the aircraft kept on the heading for which it was programmed to fly. Since we were out in the open ocean, there were no radios or routine FAA flight following calls to keep us awake like there would be on a normal training mission in the states. The last thing I remember was our co-pilot telling the pilot that he had control of the aircraft and the pilot was cleared to catch some sleep. Murphy's Law prevailed. The entire crew, including the co-pilot fell sound asleep.

We were asleep, cruising at altitude, on autopilot and heading for the Air Defense Identification Zone (ADIZ) of the east coast of the United States. Naturally, with no human intervention and only relying on the aircraft's autopilot, we missed our ADIZ entry point and control time and proceeded to fly over South Carolina. Even though this was before we were concerned about terrorists, Air Defense Command strip alert fighters were scrambled to see who and what was entering and over flying controlled US airspace without proper clearance. Both the Center and the scrambled fighters were trying to contact us on guard, but since the entire crew was sound asleep, no one aboard responded.

The pilot was the first one to wake up, open his eyes, and look out to see two armed fighters off our wing. He got on the interphone and woke up the rest of the crew. Next we established radio contact with the proper authorities, assured them that things were under control, and then ended the mission with an uneventful landing back at Seymour Johnson.

As would be expected under the circumstances, we had some explaining to do to both the Squadron and Wing Commanders. For our gallant efforts, we got a chewing out and as a bonus a good war story to tell.

The author and a different crew in a later time. (L-R) EW - Lt. Col. Larry Jensen, Nav.- Maj. Jack Warner, RN - Lt. Col. Norm Labrie, CP - Capt. Bill Hughes, Gunner - SMSgt. - John Butler, Kneeling- A/C Maj. George W. Golding. Just before a Linebacker mission!

Banned From the Officers' Club

George Golding

After a long all-night flight, three of our wing's aircraft were sent to Kinchloe AFB, Mich., (which was not a SAC base, but an ADC base), as a weather divert base. We landed about 1130 hours and after debriefing and dropping off our classified material at the Command Post, etc. we headed to the O-club to get something to eat. The lunch line had closed, so we wandered into the bar area. This was about 1530 hours. The old bartender that had been there for years was there and several ladies were sitting around two tables in the back of the room.

We asked the bartender for some drinks and snacks. He said they didn't open until 1630.

After a long conversation of us telling him of our plight, he finally agreed to get us something to eat and drink. We got the drinks first and then he went for the food. I guess we got sort of loud and this nicely dressed woman come over and informed us that SHE had the bar reserved until 1630 and asked us to leave. We sat there and said we would but not until after we ate. She was upset and said we would leave as soon as we finished our drinks. She returned to the tables. When the bartender returned with some food, we told we had been told to leave. He said that the ladies did have the bar reserved, but that if we toned it down we could stay. So we did and ordered another round. Well, Mrs. Somebody saw this and marched herself up to the bar and told us again to leave. The aircraft Commander was an old WWII veteran and he told her we were staying and having our drinks. She then said "Do you know who I am?" He said "Yes you are somebody's dependant and we are officers. Since this is an Officers' Club, we are staying right here." She went back to her friends at their tables and all the ladies got up and left the bar area.

Of course we paid for that! She was the ADC Wing Commander's spouse and as a result of our actions we were banned from the club. After a call to our wing Command Post and talking with our Wing Commander, he called the SAC Command Post and through negations, we were allowed to eat in the club kitchen but not the dining room or the bar. The excuse given was that we only had flight suits and thus were not dressed properly. Those rules applied for the remaining five days we were stuck there.

The Green Ball of Fire

Gordon Hallgren

I loved the flying part of my Air Force career far more than the 'staff' part. Let's face it, there isn't a lot of excitement to being on a staff (unless you like to get your stimulation from worrying about what the boss really wants or what kind of flak you're going to catch after not giving it to him).

I was a latecomer to the flying game, already a 1st. Lt. when I entered Undergraduate Navigator Training and a Captain when I went on to Electronic Warfare Officer Training (EWOT).

For my first assignment out of EWOT I chose the AC-130 Gunship in Southeast Asia. Now that was a thrilling assignment! But it came to an abrupt end when my SEA tour was over.

And that's when it happened. I was 'required' to have at least an obligatory tour in B-52s and was assigned to Fairchild AFB, WA, the "Resort of the North."

Getting older really stinks. I think that might be a quote from somewhere. In any case, the memory usually starts to go and details get fuzzy and all you can seem to remember is the unimportant stuff. But there were two B-52 training missions that will be forever ingrained in my mind:

Some time in about 1975, we were flying through Wilder (is that OB-10? See, I can't remember that!) on a night mission.

Everything seemed pretty normal until we departed the IP. That's when some EWOs start to sweat a little. The gunner asked me if I had anything unusual on my scope and I noted to him that both the warning receiver and the ALR-20 showed some weak and indiscriminate noise that looked like interference more than any real signals. As we completed the run and pulled out for the nav leg home, the interference went away. We thought nothing more of it and went on with our mission.

Well, it was kind of late when we landed and not too many folks were about because of the early hour, but the Chief of Bomb Nav was there to greet us! He wanted to talk to the whole crew; and wouldn't say why.

So we marched ourselves into mission debrief. The first question from him was: "Did anyone notice anything unusual during the bomb run?"

We all looked at each other kind of funny, and then the Gunner spoke up, telling him about some strobing he had had on his gunnery system. Well, of course, I had to mention the 'interference' on my systems and it turns out the navigators had had some intermittent strobing on the BNS. "OK, so what's up?" we all asked almost simultaneously.

Seems there had been some phone calls from several people in the Wilder area to the authorities. They were reporting a green ball of fire that had followed our aircraft into the bomb run, trailing above and behind for the whole run and then breaking off and climbing in the opposite direction when we exited to the nav leg.

I know there aren't a lot of believers, but just to give credence to the story I have to mention that three of the observers were the mayor, a respected local minister and the sheriff! That's my story ... and theirs too.

Northern Wedding - A Four-Ship to Europe, 1976

Lynn Wakefield

We launched on a September morning on as four-ship formation using (Minimum Interval Take-Off) MITO procedures. MITO for similar aircraft (we were four B-52D model aircraft) meant that each aircraft launched 15 seconds behind the preceding aircraft. That meant that as many as three aircraft could be on the runway in full power mode at the same time trying to achieve the same goal – to get airborne and not hit anything. What is not visible to the casual observer but is very noticeable to the aircrews in aircraft numbers two, three and in this case, four is the air turbulence created by the preceding aircraft blasting down the runway and taking off mere seconds ahead of your plane—no time for the turbulence to die down. So, Numbers two, three and four have to execute low slow turns off the original runway heading to get out of the turbulence. Because we were scheduled for a 20-hour-plus sortie, we had all loaded up on box lunches. As professional fliers we knew that MITO meant tie down the lunches well - turbulence and loose, fried chicken in the cockpit do not match well.

Flying one-mile trail formation, we were headed to Europe to participate in a combined exercise called Northern Wedding. We were going to play with several NATO countries to include the British carrier HMS Arc Royal, several F-4 aircraft from Scotland, the Netherlands, US units stationed in Germany, and the Norwegians. We were making a small footnote in history as we were the first B-52s to land in England since 1968, when the 24-hour "Chrome Dome" missions terminated.

As was standard SAC procedure, we had a full day planned - fly from Carswell AFB, Texas across the Atlantic, drop down to low level altitudes and make an interesting run across Denmark, northern Germany and into the Baltic, climb back out over the North Sea and recover at RAF Upper Heyford to debrief. Next we would meet our host unit, get a welcome briefing, local area briefing, and check into our TDY quarters – putting in about 24 hours if all goes well - just a normal SAC TDY day.

As we coasted out over Newfoundland Eastbound, all of a sudden the High Frequency (HF) radios stared to "experience reception" problems. No problem - as this was my rookie Atlantic crossing in a BUFF, I was glad to copy the traffic and pass all to the flight - turned out it was a rookie thing. They were very appreciative that I copied the SKY KING traffic every hour all night long - yeah, right. The funniest thing about that night flight over water occurred when we contacted Croughton Airways on HF so the copilot, a Georgia boy, 1st Lt. Mike "Opie" McKissack, could pass position and weather information to this fellow with a beautiful Scottish brogue. It took about ten minutes to say that the weather was perfect and that we were on flight plan. Amazingly, all the other aircraft's HF radios came back to life for that interchange!

The plan was to drop down and fly to 500' and 700' Above Ground Level (AGL) altitudes in a saw tooth trail formation during the low level portion of our flight. Aircraft number one and number three would be at 500 feet and number two and number four would be at 700 feet AGL. Each plane would station keep at one mile in trail on radar. This was a great plan expect for two things - fog and ground clutter.

We flew the entire low-level track in serious fog. We were sure glad that there were no mountains in Denmark. Because the radar returns were mostly water-land contrast and we needed higher gain to pick out the turn points along the track. This meant that the ability to see the plane ahead of us using our radar was reduced. We were talking to each other all along the track but no one could see the others or anything outside the aircraft.

After two hours of flying IFR at low altitude we finished our route and climbed out into a glorious VFR sunrise, and discovered a small surprise. That one-mile trail formation that we started over Denmark had turned into a line abreast formation and we were four miles across in a saw tooth formation. Oops! Somehow, I don't think that this information made it to the debrief session.

We were in England in 1976. At that time, Vietnam was still fresh. Ground Launched Cruise Missiles (GLCMs) were in the headlines and there was much demonstrating for a number of reasons: Pro nuclear, anti-nuclear, Pro-American, anti-American, Pan-Europe, Anti-NATO - lots of reasons to demonstrate in England in those days. So, to keep our profile low, the local leadership decided to billet all of us at a small countryside Inn some 25 miles away from RAF Upper Heyford and at least 15 miles from the nearest town of Oxford. Our crew, E-12, along with three other B-52 crews were bussed to The Belfry Hotel, somewhere in the English countryside.

After 26 hours on the road, we were glad to sleep almost anywhere. Awaking rested and hungry, most of us started scouting the place out around suppertime. Turns out that the Belfry was out the middle of nowhere, but it had a hook. The owners had turned the barn out back into a bar and dance hall called the BATS (in the Belfry). It was the best and probably only hangout for miles around. It had a rock band, cold beer, and a ton of young English women looking to have a good time, and we were in flight suits and on per diem. Life got good in a hurry.

We stayed there almost two weeks and there was not one complaint from any one. That had to be a record for any deployment. The only problem was with the flight lunches that the hotel prepared for us for the return flight to the U.S. Instead of our usual fried chicken box lunches, we got these finger sandwiches filled with butter and cucumber slices. That stuff just has no staying power on an 18-hour SAC sortie!

Nice approach, Co!

Gordon Hallgren

Now, you may wonder who the crew was on that memorable UFO mission. Actually, I left their names out for only one reason: it was the same crew that almost crashed into the runway once upon a time. Let me elucidate:

Fog was endemic to the high plateau that Fairchild AFB is on. And on one particular evening it was thick as chicken chili! Well, you know we all wanted to get home after a long flight.

The first approach didn't go so well. Some commercial guy tried to land at Fairchild in the fog, so we had to go around. Funny how having a commercial airport down the road with almost identical runway headings can generate these types of problems!

So there we were, glad to be going around because of the thick fog. Maybe there'd be better viz the next time. "Hey E Dubbya, wanna get up in the jump seat and give us another set of eyes?" Uh oh, now we knew how bad it really was.

Our co-pilot, who we'll call Fred (to protect the guilty) was due for an instrument approach. A pretty good sized, tall 240-lb. ex-football player, he was quite a sight next to our scrawny 5' 7" pilot. Crew chiefs deferred to him entirely based on his size.

Passing the middle marker and no joy on anything but fog, we motored on. We're coming down final and no lights or ground in sight. We are at minimums and nothing's visible except the illuminated white clouds of fog all around us. "Go around Fred, go around!" But Fred's hands are wrapped around the steering column with knuckles going white. Arms are locked at the elbows. "Go around Fred, go around!" Pilot tries to take control, but he's not strong enough to break Fred's grip. "Get off the controls Fred!"

Rather than dying like a bug on an interstate windshield, I elect to help out by grabbing at Fred's arms. No luck. Desperately, I raised my leg and kicked Fred over and over again. He finally let go, and just as the pilot regained control of the aircraft, we saw the runway in front of us. Which was fortunate, because we were going to land whether we wanted to or not. The pilot saved the landing and we rolled out after really getting a bumping. Tower asked for a weather observation and nearly got the truth; or a version that would at least protect the guilty.

Nice approach, Co!

Unfortunately, my two years of B-52 thrills and excitement had to come to an end and I was reassigned to RC-135's. We had many fruitful hours of droning through the skies, sucking up whatever electrons were wandering freely there, punctuated by some moments of sheer terror. But, as they say in the new all-volunteer force: "If I told you about it, I'd have to kill you."

Thunderstorm on Final

Walter J. Boyne

People often ask me to compare flying the B-47 to the B-52, and I think a fair comparison (for a bomber or a transport pilot, certainly not for a fighter pilot) would be to say that if the B-47 is a sports car, the B-52 is a SUV. It's just a little less maneuverable, and perhaps a little more difficult to fly in rough weather.

As big as the B-52 is, however, it is still a pawn to thunderstorms. When I was beginning to check out in the airplane, we went on a special high altitude mission which required us to wear pressure suits, which, as anyone who has worn them knows, are damn uncomfortable. (At least back in 1961 they were uncomfortable - they may be better now, but I doubt it.)

I was flying pilot for an aircraft commander whom I won't name, out of respect, for he was truly one of the best pilots I've ever known. (Let's call him "Buster.") Buster was envied by all (including me) because he was the real rarity, a "natural pilot." Flying was effortless for him, he anticipated everything, and it was a joy to watch him.

But he did tend to be a little irascible and short tempered, and this quality had been exacerbated by our relatively short, four hour flight in pressure suits. We were returning to Kirtland on a summer day, with summer storms skirting through the area. You could clearly see individual thunderstorms moving through what was largely clear, if bumpy air.

He told me to call for a straight-in approach, which was reasonable, as we were short on fuel (you couldn't get to pressure suit altitudes in our B-52C if you had any sort of a fuel load.) As we continued our approach, a thunderstorm was moving from north to south, across the west end of the field, about a half-mile out from the runway. We got the usual warnings from ground control, and closer in, from the tower, but I didn't worry about it, for it would either clear out of the way by the time we got there, or we would go break off and make another approach.

To my amazed horror, the A/C kept boring in. We'd completed the check list, and there was no one ahead of us on the runway, but the thunderstorm had stopped right off the end. I looked over at him, waiting for him to do something.

There is no doubt in my mind today that I should not have watched, but instead, grabbed the control column and made a ninety-degree turn. Instead, I stupidly watched as our aircraft plunged right into the thunderstorm.

We didn't go far. That storm spat us out in a vertical bank to the left, roaring over the hangar where base operations' aircraft were kept. "Buster" didn't say a word, just manhandled the airplane back to straight and level, made another 90 degree turn and re-entered the pattern. By this time the thunderstorm had moved off, and he made his usual smooth touch down. Not a word was exchanged as we left the cockpit, and no one on the ground, knowing Buster's temper, mentioned it either. But no one on the aircraft ever forgot it; there's something about looking out the side window, straight up, in a B-52 that is memorable.

Guarding Our Freedom –

The Cold War

President Gerald Ford visits the Carswell Alert Force.

Alert crew badge, dog tag, and P-38 can opener.

The B-52 and Nuclear Alert

Walter J. Boyne

The Strategic Air Command was without question the most influential component of the United States' armed forces, and it routinely received a large share of the military budget. But even a large budget and good management could not solve all the problems, and the usual last resort was to take a solution out of the hides of the crews. The alert system was one of these last resorts, intended to solve the problem of maintaining a large proportion of the force instantly ready for war.

While the alert system was undeniably effective, it was also cruel to family life. The ordinary working week averaged sixty to eighty hours, and there were lots of Temporary Duty (TDY) separations. The alert system added to the hardships. Sometimes the stress of life in SAC became too much and divorces resulted, but for the most part, the wives and children pulled their "tours of duty" with the same courage as did the crew-members.

Under the alert concept introduced on October 1, 1957, the Boeing B-52s were required to get airborne within fifteen minutes of the order to take off. At the time, it was widely accepted that the U.S.S.R. had a great advantage in ballistic missiles that could strike U.S. targets thirty minutes after being fired. With immediate warning, a fifteen-minute alert meant that the B-52s would just have time to get off to retaliate for the missile attack.

Crews on alert status were expected to remain together and in close enough proximity to the aircraft to make the fifteen-minute schedule. Aircraft were "cocked", i.e. ready for engine start, and experienced crews could have the engines running within two minutes after the alarm and be taxiing within five minutes.

One of the unacknowledged problems of the alert was unique in warfare for an enemy attack meant that the families were more in danger than were the warriors. If an actual enemy ICBM attack was imminent, the crews would takeoff to retaliate. The chances were that they would return from the combat mission, only to find that they had lost their families to the enemy attack. There were evacuation plans but Civil Defense was never really embraced by the United States as it had been in the Soviet Union. Tension remained high throughout the Cold War despite the frequency of training alerts, for the crews never knew which one might be the real thing.

The routine and schedule of alert duty varied from base to base and over time, but a crew might expect to pull one-week of alert out of the month, in addition to all their other many training requirements. In the early days, the "alert shacks" were improvised, but later they were well built and relatively comfortable. In their "free" moments the crews could use the usually Spartan recreational facilities or work on correspondence courses, but much of the time was spent in studying their Positive Control Procedures and their specific war plan mission The crews had to be as concerned about Positive Control Procedures and the associated paperwork as they were about the war plan mission, for any failure with the procedures—or loss of the paperwork—meant serious disciplinary action.

The B-52 did not go to war for more than ten years after it entered service but its entry into the Vietnam War would have a drastic impact on SAC's ability to maintain a substantial percentage of its force on alert.

Crews leaving for morning preflight from the Carswell AFB Alert Facility. Notice the down-ramp from the center of the building and the two up-ramps on the sides leading from the sleeping area

Sitting Alert – A Personal View

Tommy Towery

The state of "Nuclear Alert" is a complicated and extremely complex subject to try to explain to civilians. They have never been burdened by the obligation placed upon the shoulders of those who have lived that life. To get even an idea of the big picture is like trying to nail jelly to a tree. Just getting a general idea of the subject requires understanding the activities, the facilities, the mental state, and the commitment of the personnel involved.

For most of the Sixties and up until 1991, up to one-third of the Strategic Air Command's bomber and tanker forces were normally ready to launch for a nuclear war within fifteen minutes of when the signal sounded. This manned bomber threat was accompanied by supporting tanker aircraft and crews, along with land and sea launchable nuclear missiles acting as a deterrent to any country that might decide to begin a nuclear war with the United Stated. The Nuclear Triad force commanders saw the manned-bomber force as a way to show intent, without actually launching a strike that could not be recalled once it was started. If threatened, the bomber force could be sent airborne and orbit at designated areas, which would make them less venerable to destruction on the ground, as well as place them closer to their own targets.

In trying to explain this phase of SAC life, I would like to draw an analogy to the honorable game of basketball. Depending upon when and where you were born and grew up, the rules of basketball were not always the same. In the early days most people played a half-court game; later, full court was the game. College and pro basketball courts are laid out differently. Today there is a three-point shot – in the earlier years, there was not. Lines painted on the court change, period timings change, clocks change and uniforms change constantly in the simple game, but the game's still called basketball. Take a complex issue such as nuclear alert (along with the rules applied to it), and you find that there are just as many variables in the game of "thermo-nuclear war" as there are in basketball. For the most part, the observations on alert life that I describe are based on the way the game was played when I was a player. Specifically, I was assigned to B-52D models at Carswell AFB, Texas, from 1971 to 1978. My memories are based upon that setting and with the game rules that were enforced at that time. Those that played before or after me will have stories that may vary from mine, but it was still the same basic game. I came after the "Chrome Dome" days of airborne nuclear alert and the 24-hour missions, which that concept demanded – thank goodness.

My alert life cycle began on a Thursday morning when a large number of Crewdogs would tell their girlfriends or wives and families goodbye, and load up their gear in their cars and trucks and head toward the base. At a scheduled time they would arrive at the Carswell AFB alert facility to relieve the Crewdogs that had done the same procedure only a week before. Some would carpool with other members of their crew, and some drove themselves. Those who had only one car for the family would be driven to the gate by a wife who had to get up and get herself and the kids dressed so that she could have the car at home for the next week. Each Crewdog knew that once the Security Police checked their badges and they walked inside the guarded gates leading to the facility, they would not be allowed to leave the tether of the alert grip for the next seven days, no matter what happened in their lives. Once they had accepted the aircraft and the mission of the crew they relieved, they signed away their freedom for a full week. It could not be more restrictive if they were being put in jail, even though the barbed-wire fence around them was designed to keep people out and not in. Most of the Crewdogs I knew pulled an average of 120 days a year on alert after the bombers and tankers returned from Southeast Asia and ground alert was renewed.

Alert was a big part of life for early SAC crews. At the time that many of us Crewdogs pulled our alert tours, it never seemed like there would ever come a time in our lives when we felt a need to document the living conditions of the facility for history. Now, I feel differently about that need.

During the timeframe from 1971 to 1978, when I routinely was required to occupy a position in that facility, there were two basic things in which I believed. The number one belief was that we would never actually get the call for which we were waiting to launch our aircraft and participate in a nuclear war. We called ourselves prepared for it, and on paper and in our hearts we were. Had the horn gone off, we would have launched our aircraft, and we would have done what we were trained to do; however, no one ever actually thought we would become involved in a nuclear war or have to drop our weapons on enemy targets, or kill millions of people. That was one of the major reasons we were standing alert to begin with – as a deterrent to such possibilities. If our enemies knew and believed that we had the means and the will to retaliate for an attack on us with such destructive power that their country would sustain major destruction in return, then the option for them to strike us first with a nuclear weapon did not make sense.

The number two truth that I felt was that there would always be crews on nuclear alert. The idea that someday the U.S. might be the only nation with first strike capability was inconceivable. We felt that we would always have a nuclear threat hanging over our heads. We were a generation that grew up with duck-and-cover drills in school and with families who built fallout shelters in their basements. We had replaced another generation of crews who felt the same thing, and we thought that the next generation that relieved us of that obligation would feel the same threat to them and their families.

With these beliefs at heart, it seemed silly at the time to actually document the alert facility and the day-to-day life of an alert tour, since there would always be Crewdogs pulling them. It is one thing to tell war stories, but even though this period of time was called the Cold War, it did not seem interesting enough or important enough to commit to words. Things change, and in 1991 SAC crews came off of alert one day and no crew took their place. Alert was over. It was history.

Most bomber bases had their alert shack, as it was often affectionately called, at a distant end of the runway. It and the bomb storage facility were the most restricted places on base. Near this facility was the "Christmas Tree" parking ramp where all the alert aircraft were placed. This placement was based upon providing a short taxi distance for the bombers and tankers to get to the runway and launch in the fastest time possible. My first awareness of concertina wire was when I saw my first alert shack. The tall fences that surrounded the facilities were topped with this razor edged wire that was coiled in big rolls like barbed wire atop the fences.

Entry into the alert area was through a controlled gate manned by armed Security Policemen. These professionals took their jobs very seriously. A red rope was strung a safe distance around each plane and entry into the planes in the enclosed area was under the control of a two-man policy requirement, which was enforced by the armed guards. No one person was allowed to be near the plane alone, not even the wing commander. The answer to the most commonly asked question about the planes parked there was the book answer of, "I can neither confirm nor deny the existence of nuclear weapons aboard this plane." Behind each of the big bombers was a blast fence made of metal sheets that were curled upwards to deflect the thrust of the engines. It was behind these blast fences that crews were required to park their trucks during an exercise drill. They could not allow empty trucks to sit beside the planes when those big babies were being taxied into position. For that reason, the gunner normally drove the alert truck when crews went places together. When responding to a klaxon, he would drop the front-end crew off and then drive to the rear of the aircraft where his crew entry hatch was located.

The normal alert facility building was two stories high, with the first story being underground. The bottom floor contained the crew sleeping and shower room area. It was constructed of cinderblock walls and since it was underground, there were no windows to the outside world. Depending upon the number of crews assigned and the level of national alert (DEFCON), the sleeping rooms could house from one to four Crewdogs. The showers were common areas and included the toilet facilities as well.

There were inside stairs leading to the top level of the alert shack, and there were concrete ramps from the outside that led from the lower level to ground level, as well. These ramps were designed to provide quick exiting of the building during an alert when the crews had to run to man the planes. Likewise, there were ramps from the top level to the ground on all four sides of the building that served the same purpose. When five bomber crews and five tanker crews are all sitting around engaged in activities ranging from taking a shower to eating a meal, one can only imagine the level of activity when the klaxon blew. The last thing needed was for someone to fall down a flight of stairs when the future of the nation depended upon the alert aircraft being launched on time. It was much easier to run up or down a ramp than to try to negotiate a set of stairs with twenty people hot on your heels.

The major part of the daily activities took place on the upper level. Just inside the front door was the office where crews checked in and out. This was where the handheld radios were issued to crews leaving the area. The rule was that when the battery went dead on a handheld unit, you had to return to the facility. There was a hallway to the left past the office that led to mission planning and study rooms. A hallway to the right led to the snack machines and a small office that was visited by the flight surgeon during his scheduled hours. There were also pay phones there. This was before the time of cell phones, so the whole group had to share the pay phones for personal calls off base.

A central hallway took you to a small library on the left past the stairwell, and a TV viewing area to the right. There was one TV and the station to watch was determined by force or majority vote. A little further down and to the left was another hallway that took you to an area that had a pool table and ping pong table. You didn't want to get the sleeping room under the pool room because of the way the balls kept getting knocked off the table and bouncing on the tile floors.

The last room on the right of the central hall was the chow hall, where alert crews had their meals served. Meals were not free to the officers on alert. Each meal was paid for individually, and for many years was an all-you-can-eat buffet style meal. Later the meal plan went to an ala-carte plan and that broke the bank for a lot of heavy eaters, especially for my gunner who would consume a dozen eggs for breakfast.

The last room on the left side of the central hallway was the briefing room, which doubled as the theater in the evenings. There was a swimming pool just outside the facility, inside the guarded fence, that was available during warm weather and was used primarily as a place to throw Crewdogs pulling their last alert tour before retirement or assignment or giving up on their idea of an Air Force career.

A normal tour began with the new crews all assembling in the briefing room. On changeover morning the outgoing crews went straight to the aircraft to make arrangements to take their own personal equipment off as soon as relieved. A quick roll call was accomplished to insure that each crew had all members present and that they were fully manned to relieve the crew going off their tour. Next, members of the wing staff gave them weather reports and an intelligence update.

When released from the briefing, the new crews went to the alert aircraft to place their own gear aboard and start a preflight of the aircraft. Inside the aircraft the Electronic Warfare Officers and the aircraft commanders of the two crews did an inventory of the combat mission folders and the Emergency War Order "tickets". The radar navigator did an inspection of the weapon's area, and the gunner checked his compartment and equipment. Various safety and two-man policy procedures were done and finally the new crew signed for the mission, while the old crew was released from their alert duties. At that point, if the klaxon sounded, the new crew had the mission responsibility.

The old crew quickly gathered their personal and professional gear and headed to the parking lot. There was almost always at least one car that required jumper cables to be started after sitting up for the previous week. Once relieved, the crew coming off alert entered a condition called "C-Square," which was short for "C-Square, R-Square" (Combat Crew Rest and Recuperation). It would be 8am on Monday before they had to be back on base. They would have the rest of Thursday, all day Friday, Saturday, and Sunday, to do with as they pleased. While this seemed like a generous period of time off, it really only made up for the weekend they had missed before by being on alert and the current weekend, but all the days were strung together.

As for as the new crews that assumed alert, they had the next week of "captivity" to deal with. It would be impossible to try to detail all the activities that were awaiting them. Many training classes were completed, simulator sorties logged, and emergency procedures studied. For some, it was a time to work on Squadron Officer School or Air War College correspondence courses. Others engaged in college level work for whatever college or university that had the contract at the time with base education.

Most of the weekdays were filled in the same manner and the same activities that the Crewdogs would be doing if they were not on alert. If there were a planned mission the following week, the mission planning would be accomplished. Flight Safety briefings and intelligence briefings were given to the captive audience. Varied types of additional duties, such as Combined Federal Campaign work and Suggestion Officer paper work were also accomplished.

When no official activity was scheduled during the day, watching TV and taking naps filled the long hours. At Carswell we had different classification naps. A Class-C nap was flight suit and boots on and sleeping on top of the covers. A Class-B nap was boots off and flight suit optional, but again, atop the covers of the bunks. A Class-A nap was flight suit and boots off and under the covers with the lights out. Of course, these were not official names.

Many trips were made to the Base Exchange (BX), and crews normally traveled outside the gates together. The hospital at Carswell was outside the main base gate, so that was off limits. That required the flight surgeon to make a daily visit in the mornings to care for the minor medical problems of alert personnel. It also meant that if a Crewdog's pregnant wife selected the week he was on alert to have the baby, he was not available to be at her side. The squadron schedulers took this into consideration, but due to the unpredictability of due dates, many a new father found out by telephone.

Travel to anywhere on base that did not have a klaxon horn installed or any facility that was so far away that a 15-minute launch response time could not be guaranteed was also not an option. Some hobby shops were okay, but many were in isolated parts of the base and so were restricted.

It should go without saying that alcohol was strictly prohibited for the entire week that crews were on alert. For some odd reason SAC commanders did not want B-52 aircraft that might be loaded with nuclear bombs being flown by drunken crewmembers. Nor did they want flying gas stations like the KC-135 to go screaming down the runway for a takeoff under the control of a person who would be considered DUI if it was a car he was driving and not an aircraft. So there was no drinking allowed, not even wine with a meal. As far as meals go, most of them were eaten in the dining facility in the alert shack because of the convenience and not necessarily the menu. On special occasions, the choice for some meals was the Officer's Club, NCO Club, or the bowling alley. After dinner, the evenings were filled with watching TV and other non-military things. This is where personal preference came into play the most.

The age in which I pulled alert was before cell phones, laptop computers, iPods, and even video tape recorders. There were not even Walkman cassette tape music players around. It was pre-cable TV at Carswell and so the one TV in the TV room offered only the three local network stations (ABC, CBS, and NBC) and public television. This was probably a blessing in disguise, since there were not hundreds of channels from which a group had to agree upon to watch, only three or four. Ironically, our public television station got a lot of air time for the alert crew, especially when Benny Hill and Monty Python became popular. It is difficult to imagine how it would ever be decided what program to watch on cable TV with 30 or 40 guys trying to vote on the stations. Sports and movies were normally the TV entertainment of choice.

Since there were no VCRs or DVD movies, the alert crew was treated to movies sent in by the Armed Forces Motion Picture Service, or some group like that which supplied movies to remote units and base theaters. Each week new waves of 16mm film cans, which contained pre-selected movies, were delivered to the alert facility. They were shown on a projector in the briefing room, and the gunners usually had the skill and talent to run the projectors. Since there was only one projector, there were intermissions whenever the reels had to be changed out. At my base the gunners also had another trick up their sleeves. They had a set of dirty movies. I don't know where they came from, who owned them, or how old they were, but periodically we'd show the same old stag films in a room protected from the outside world by barbed wire and armed guards at government expense.

The normal selection of acceptable movies was usually the type shown at the base theaters, but normally were a few months behind in their release. There were times when many of us watched movies without having a clue as to why we were sitting there, but it made the captive time go faster it seemed. A new rating system came to be thanks to my alert tours. If a movie was bad enough to get up and walk out of while pulling alert, it had to be one of the worst movies ever made. Otherwise we sat and watched almost anything that was shown on the screen. More often than not, the lines supplied for the onscreen action that came from the audience were better than the ones written in Hollywood. I am sure that "*Mystery Science Theater 3000*" writers got the idea for their show's format from a former alert crew member.

Card games were also an easy way to waste a lot of evening time and a lot of money. While many played non-betting games such as Hearts or Spades or Bridge, poker was the king of the hill in card games for alert crews. And with dealer-calls-the-game as a standard rule, it was a varied set of skill levels needed to keep from loosing you shirt. Some with memorable wild cards were "Whores, Fours, and Buggy Tops" (queens, fours, and sevens), "Follow the Queen", "Suicide Kings", "One-eyed Jacks" and even a few hands of "Indian" when things got really crazy. It's hard to picture a bunch of trained, hardened, combat-ready warriors betting money while holding one card on their foreheads like a single feathered headdress.

Yes, they played with real money and some lost big and some won big. While it was usually a quarter a bet and three-raise maximum rules, the pots could still get high stakes in a hurry, if you had seven players. The most dangerous games were the match-the-pot games, and in particular the most dangerous game at Carswell was a Cajun poker game called bouree (Booray). It was similar to a high stakes, single hand bridge game. I don't know if it migrated to the northern bases or not, but in Texas it was quite popular. It was dangerous and expensive because if you made any mistake in protocol such as turning over your trick when you dragged it from the table, you had to match the money in the pot. If you did not take a trick, you matched the pot. If the person that played before you made an error and had to match a $4.00 pot, the pot was then $8.00. If you made a mistake on your turn, you matched the $8.00 pot making it $16.00. If you started a game with seven players and there were only five tricks to be taken, someone almost always had to match the pot early. Some people dropped out of the game not because they had bad hands but because they were afraid of making mistakes or getting burned. You burned someone by keeping him from taking a trick. It was played until there was only one winner for the pot, never split. And it was played with what was called

"Combat Rules" which meant that you had to know the rules before you sat down, because there was no leniency when it came to beginners making mistakes. It was the one game that drew more spectators than players, especially as the pot grew. Make a mistake and you match the pot. Booray!

Another popular activity during my alert era was playing Avalon Hill or Strategy and Tactics war games. This was before the days of computer games so we used little cardboard pieces on an outstretched map and rolled dice to determine the outcome of battles. Full scale campaigns could take days so it was a great time killing activity. Some Crewdogs built model airplanes and cars, some worked on Heathkit or similar electronic projects. I walked downstairs one night to find a full skydiving parachute stretched out in the hall and a gunner repacking it. I took my lead bullet molds one tour and like Tom Sawyer and the fence painting, had about a dozen fellow Crewdogs fighting to cast bullets for my black powder rifle. One of my friends used the time to design and draw out the plans for a house he built.

Things were so boring on one remote alert tour that my husky, Oklahoma-farm-hand co-pilot taught the whole crew how to knit watch caps, and we all went to the base exchange and bought knitting needles and packages of yarn of various colors and sat watching football games on TV while we knitted. I had a gunner who brought his haircutting supplies and would give free haircuts to anyone who took the risk. We Crewdogs did what they had to do to pass the time.

On very rare special occasions, families were allowed to come out to the facility – the one time I remember was on Christmas Day. When I started pulling alert, there was no place for families to visit, except for the base recreation center or the picnic grounds in warmer weather. In the mid-Seventies a Family Visitation Center was constructed near the alert facility, but outside the gates of the alert area. This building had several rooms, and visitors could come out and have some privacy with the alert crews. It was usually filled with screaming kids running amuck while the parents visited with each other ignoring them. For that reason, many Crewdogs elected to remain in the sanctity of the barbed-wire-protected alert facility.

When you describe the features and offerings of the alert facility, it really doesn't seem so bad in retrospect. Here we had an exclusive club, which until the mid-Seventies, was a "Men Only" club. The members had a free place to stay overnight with maid service, and a place to eat three meals a day cheaply. There was a library with all the current magazines, a TV viewing room, a swimming pool on many bases, a game room, a theater with free movies, and you had free use of a company vehicle. You could not and did not have to leave to go home and work on the car, or hang pictures, or mow the grass or shovel the snow. You did not have to even drive to work. You did not have to worry with putting kids to bed or getting them up and off to school. You did not have to worry about what to put on in the morning since it was always a flight suit. Every night you either could play poker with your buddies, or go to the gym and shoot hoops, or hang out in the pool room playing eight-ball. On holidays you sat around with all your pals and watched sports as long as you wished. When you went to the Base Exchange you could walk to the front of the line (no matter how many people were already standing there). Alert Crewdogs had top priority all the time!

That was life at the alert shack. Put this baby on a mountain top or on an ocean liner and you've got a vacation to die for, but therein lay the catch. This was an exclusive club. You had to be willing to die to become a member of this club, literally, and not just in a figure of speech. You had to be willing to stop whatever you were doing, run out to an airplane in rain, sleet, snow, or blistering heat at any time day or night and climb aboard it. You had to be willing to leave your loved ones behind at any given moment and fly off into combat of a type the world has never seen before and into a war that may not have a winner. You had to accept that if that time ever came, you might never see your family again. You had to be willing and able to do that by instinct, without thought or hesitation, and even if you were willing to die yourself without hesitation, you also had to be willing to kill hundreds of thousands of other human beings just as well. You had to spend countless hours to prepare for membership in that club. You had to sacrifice days, weeks, and months of your life just to pay your dues to stay a member. You even had to give up personal habits, needs, and desires; and if you were willing to accept all that, then you were allowed to be locked up for a week at a time and call yourself a member of "The SAC Alert Force."

The author's 1974 promotion folder photo.

DEFCON 2 Cuban Missile Crisis

Arthur Craig Mizner

From 1956 to 1976 I was a professional Strategic Air Command (SAC) flight crewmember in B-47A/E and B-52 B/C/D/E and F. My position was that of a co-pilot, pilot, aircraft commander, and instructor pilot. During those years I logged over 11,000 military hours and 113 combat missions over highly defended SAM/AAA targets (Linebacker I/II) plus many other combat missions in South East Asia (SEA). In addition, I flew over 100 24-hour-plus airborne nuclear alerts missions called Head Start, Wire Brush, or Chrome Dome. I also spent over 10 years pulling ground nuclear alert at my home base and at forward echelon bases. The many years I spent as Squadron Instructor Pilot (IP) and Replacement Training Unit (RTU) IP added a few more white hairs. However, there is a 27-day at Defense Condition (DEFCON) 2 that stands out as the most important period of my United States Air Force (USAF) career. That period was the Cuban Missile Crises (CMC) of October 1962. The CMC threaten the entire free world and I became one of the key players in the USAF response to halt the deployment of Intercontinental Ballistics Nuclear Armed Missiles (ICBM)s from the Soviet Union to Cuba. This period of 27 days is regarded by most military leaders as the moment when the post Word War (WW) II Cold War was the closest to becoming a nuclear WW III.

My first ground alert started in late 1956 in B-47Es, loaded with atomic MK-6 Mod 6 weapons, initially at my home base then followed by forward echelon bases in 1957 and 1958. The forward echelon bases were located in Alaska, England, Guam, North Africa, and Spain.

Since May 1960, SAC kept more than 400 B-47, B-52, and B-58 bombers and supporting KC-97 and KC-135 tankers on 15-minute ground nuclear alert at home base and forward echelon bases. In addition, several B-52 bases started flying 24-hour nuclear airborne alert missions in early 1961. That year, my crew flew over 50 airborne nuclear alerts by the year's end. We did not pull ground nuclear alert during this period of airborne nuclear alert.

In October 1962, I was a USAF Captain with 2,800 hours, (1,000-plus hours in B-47 A/E, and 1,500 hours in B-52 B/F) flying B-52F aircraft stationed at Columbus AFB, MS, 4228 Strategic Wing (SW), 492 Bomb Squadron (BS). My position was that of a senior Co-Pilot (CP) but I was on the upgrade list to Aircraft Commander (AC). The requirement for a B-52 AC was 2,000 hours in the B-52 unless you were an AC in another SAC bomber aircraft i.e., B-36/B-47/B-58.

The below events are from part of an article I helped develop for the November 2005 issue of the *Smithsonian's Air & Space Magazine.* For additional details, please read the article *Cuban Missile Crisis 27 days at DEFCON 2.*

Before the President of the United States of America (USA) told the public about the CMC, the entire USA military was placed on full ground alert. On 22 October 1962 I was on ground nuclear alert in the alert briefing room with all the other ground and flight crewmembers watching President Kennedy's televised speech. The President told the American public about U-2 reconnaissance flights finding SAM and ICBM sites in Cuba. The Soviet Union was preparing to deploy medium-range R-12 missiles to Cuba. The missiles had 2.3-megation warheads and a 1,100-mile range. The R-12 missiles could reach Philadelphia, Saint Louis, Oklahoma City, San Antonio, and the Panama Canal. My alert base of Columbus AFB, MS was well within the missile range and a prime first-strike target. In addition, photo-interpreters discovered that Soviet technicians were also preparing sites for 16 more R-14 ICBMs with a range of 2,300 miles. After the President's speech, no one said a word for about five minutes. Most of the flight crewmembers had seen aerial combat in WWII or Korea or both and these combat veterans understood the impact of what a nuclear WWIII would be like.

That night the US military went on DEFCON 3 alert. Aircraft that were undergoing modification at Depot Level were put back into flying condition and returned to their home base. At SAC bases around the world, flight crews and ground crews raced to get every flyable bomber and tanker fully loaded with fuel, ammunition, and nuclear bombs to a cocked position. A cocked aircraft is one that has been fully serviced, pre flight preformed to include running engines and all system and then shutting down. The fuel was topped off and all switches were positioned up to the before starting engines checklist. All target material and personal equipment was in place on board the cocked aircraft. All coded messages (GO CODE) were kept with an armed crewmember. On 24 October the military went on DEFCON 2 alert.

The 4228 SW was composed of 15 B-52F and 15 KC-135A aircraft. All tactical aircraft and all ground and flight crews were on full ground alert. The only flying done was to perform functional check flights after major aircraft repairs or to ferry aircraft from the depot to home base. No routine ground training was scheduled. After President Kennedy's speech, two B-52F carrying four B-28 FI thermonuclear bombs started flying 24-hours-plus airborne nuclear alert. The base was restricted to military personnel only.

Other B-52/KC-135 SAC bases were on the same type of alert but their ground alert and airborne alert supporting schedule may have been different. Their bomb loads could be of different types of nuclear bombs up to include the mother of all bombs, the 25-megation MK 41 Hydrogen bomb. The 4228 SW had several schedules but the one that seemed to work best was three days on ground nuclear alert followed by airborne nuclear alert. During the first three days on ground nuclear alert, the B-52 flight crews would accomplish airborne nuclear alert mission planning including fail safe items, target study and questioning and certification by the senior wing staff. The routine also covered survival training, including safe area in case of bail out, mission briefing, and daily intelligence briefing. All crewmembers were restricted to the alert compound but were allowed to meet their families at the alert facility main gate for a short period-of-time. The out going flight crew would transfer control of the ground alert to the on coming flight crew at 7:00 AM. The out going flight crew would then attend an Airborne Alert Pre-Takeoff briefing followed by aircraft launch at 9:00

AM for 24-plus-hours nuclear airborne alert. After landing the next day after 9:00 AM, we had the rest of the day off for crew rest and then back on ground nuclear alert at 7:00 AM. My crew flew five airborne nuclear alerts and I flew one extra one because one of the co-pilots was on a no fly status (DNIF) and he sat my ground alert as I flew his airborne alert.

The mission profile for the 4228SW was to fly east out of Columbus AFB, MS with an in-flight refueling (110,000 pounds) near the East Coast. We could have either two KC-97s or one KC-135A depending on availability of tankers. The next in-flight refueling (128,000 pounds) was as we entered Portuguese airspace. An orbit pattern was set up in the Mediterranean Sea as far east as Crete. We departed the orbit pattern to return-to-base (RTB) when our fuel status reached a pre-determined value. The crewmembers would take turns monitoring the HF radio for SAC traffic or a GO code. The GO code would be transmitted from SAC headquarter if the President ordered a nuclear strike. Some crewmembers reported MIG 17s flying off their south side as they passed north of Libya. On one mission while flying north of Libya, I saw two jet fighters flying parallel with us off about five miles south but we could not make an identification (ID) on them. The Electronic Warfare officer (EWO) reported being scanned by radar out of Libya. While we were in our orbit pattern, we were within two hours of several assigned targets. The targets depended on our position in the orbit pattern when we received the GO CODE.

All crewmembers had to be strapped in their ejection seats during takeoff, in-flight refueling, in-flight emergencies, and landing. During the flight, the crew could get out of their ejection seats for a short rest. Each compartment (upper, lower, and tail) had to have a crewmember on aircraft oxygen at all times. The crewmember out of their ejection seat had to be on aircraft oxygen using an extended oxygen hose or a have a portable oxygen bottle attached to their oxygen mask hose. The oxygen mask did not have to be in place.

From time to time an extra pilot or navigator would be on board to fill in while the primary crewmember relaxed. In most cases, these extra crewmembers would get in the way or cause additional stress. These former crewmembers had been sitting at a desk for years at a headquarters level position. On one of my missions, a Lt/Col pilot type from the Pentagon came along to observe the entire airborne nuclear alert from Pre-Takeoff briefing to Post-Mission Maintenance and

Operational debriefing. Based on some of his recommendations, several local in-flight food preparation procedures were changed for the better. During the flight, I asked the Colonel to put my frozen in-flight dinner into the convention oven as I was coming on my break period. The frozen dinner was provided by the base in-flight kitchen. The dinner was a frozen steak on a plastic high temperature plate placed inside a card board box. The instruction stated, "Remove contents from the box and place in the oven for 20 minutes." The Colonel put the cardboard box and contents into the convention oven and within five minutes the cardboard box caught fire filling the cockpit with smoke. The crew had to go on 100 % oxygen with masks in place. This smoke condition lasted for about 20 minutes.

The crew comfort was very poor from the lower compartment freezing temperatures to the sun-heated upper pilot's compartment. The only position where a crewmember could stand up was on the ladder coming from the lower compartment to the upper compartment. The lower compartment had no space where a crewmember could lay down. The upper compartment had some floor space but with all the cold weather flight gear and extra parachutes resting space was very limited. There was a urinal and commode in the lower compartment and a liquid heater (hot cup). A convention oven was located behind the AC's ejection seat. The target material container was located beside the oven. The aft compartment (gunner's) had two seat positions, either full up or full down. The down position was designed to enter or exit the aft compartment. The down position could also be used for resting. In the down position the gunner could not see out of the aircraft and was unable to help the pilot team look for aircraft.

During the mission, the crews flying the missions to the Mediterranean Sea and orbit over the water were asked to look visually and with radar for ship convoys heading toward Cuba. All sightings were to be reported to SAC Headquarters. Our crew made some sighting reports but never heard any results if they were Soviet ships.

I later read that there were 72 nuclear loaded bombers on airborne alert during this period with 1/3 of them within two hours of their targets. The other 2/3 would be either approaching the orbit pattern or RTB. In either case, if a GO CODE was received and they had enough fuel to reach their targets before dry fuel tanks, they would strike their targets.

Most all the bomber crewmembers felt the mission was one way with very little hope of returning home. The families living near a SAC base or prime military felt they had little hope of survival either by the first strike or radiation poising.

After the two 9:00AM B-52 launches, the returning B-52s would land. Some Wing Commanders would have the pilot make a low approach followed by a radar controlled pattern to landing. The purpose for doing this was to adjust their depth perception to low altitude flying. If one of the B-52 could not get airborne, the returning B-52 was to perform an in-flight refueling and stay airborne until a replacement B-52 could get airborne. At my home base, this did not happen.

During the 27 days of DECCON 2, SAC crews flew 2,088 sorties. Two RB-47s crashed on takeoff killing seven and a U-2 was shot down over Cuba killing the pilot. These men plus all the men and women of the military supported by their families avoided a nuclear WWIII with their show of might.

The CMC changed my life and that of all military personnel in a positive sense. Our response to the events showed the free world that the USA stood behind the goal of freedom for the people. At my age of 70 plus, I still can remember most of the events and see in my mind the aerial displays of the many missions.

As a reward for my support of the CMC, I was able to attend the 14-week Squadron Officer School (SOS) Class 63A starting in January 1963 at Maxwell AFB, AL. I had taken SOS by correspondence while on ground nuclear alert. In the past, SAC could not spare the time for combat ready crewmembers to go to SOS in residence. SOS had a positive change in my military career and my life.

The CMS provided me with my understanding of the inter strengths that helped me through my many missions in Viet Nam. We knew that there was no return for a nuclear strike into the Soviet Union compared to the feeling of returning from missions in Viet Nam.

After the return of all ICBMs from Cuba back to the Soviet Union, SAC went back to normal alert posture of 1/3 of the force on ground and airborne nuclear alert. I would continue to fly 24-hour airborne alerts for the next eight years but at a level of about 10 per year.

When I retired on 31 December 1976 my Form-5 showed I had logged 9206.7 hours in the B-52. The B-52 was my home away from home. I also had the same number of landings as I had takeoffs so the B-52 took care of me.

The author sitting at his EWO crew position.

Satellite Alert and SAC Security

Gary Henley

The 7th BW (Bombardment Wing), Heavy, at Carswell AFB, Texas, was called a "Super Wing" due to the huge number of B-52Ds and KC-135s on the ramp. This included large numbers of aircrews, maintenance, security, and support personnel required for two bomber squadrons, a tanker squadron, CCTS (Combat Crew Training School), and CFIC (Consolidated Flight Instructor Course). A *lot* of peacetime training was accomplished at Carswell; however, a lot of Alert was pulled by the crewdogs at Carswell, too. In an effort to assure a better chance of survivability in a nuclear war, an additional number (sensitive) of planes were first dispersed to the old Amarillo AFB, Texas, and then later to Sheppard AFB, Texas, near Wichita Falls. Fresh new set of crews were sent there weekly to sit alert in a SAC alert facility located at the austere part of the base, just off the end of the runway.

Sheppard AFB was an ATC (Air Training Command) base that hosted mostly medical technical training and UPT (Undergraduate Pilot Training) units, and this lone SAC detachment was a tenant unit. SAC was allowed to use the old alert facility, but only the bottom level. It had to be modified somewhat to remove some of the sleeping quarters and replace them with a kitchen/dinning room and TV lounge. The upper level had been converted to office space years earlier. Since the alert facility was located too far away from the rest of the base, and since alert crews could not possibly respond to their alert aircraft within the SAC required response time, all SAC aircrew, maintenance, and security personnel were confined to a small alert facility area for the entire duration of each satellite alert tour. We were all "in the same boat." Bottom line: the only thing we ever saw at Sheppard AFB was the alert facility, our aircraft each morning, and the ramp. The only deviation from that was when we might fly a B-52 into Sheppard the day before to swap out an aircraft or when we flew the swapped-out B-52 for a FSAGA (First Sortie After Ground Alert) training mission, ending back at Carswell AFB. [Note: At least, that's what we *thought*, given our limited experience sitting satellite alert. This was relatively new for us in early 1976, and crewdogs would groan when they heard they were going to Sheppard.]

The Det (detachment) commander at Sheppard ran a "tight ship;" however, his no-nonsense approach was balanced by a vigilant eye on the morale of the troops. He ensured that we had the latest movies from base to watch, he dropped hints about the no-notice klaxons and exercises (e.g. "If you start a movie before 3PM, you might not be able to finish it today without interruption."), and he let us vent a little steam once in a while by pulling pranks on each other and by initiating new crewdogs to alert life. Naturally, since Wichita Falls is in the middle of "Tornado Alley," it was important that rookies spend some time on the roof of the alert facility, pulling "Tornado Alert duty," right? The Det commander would look the other way and pretend that he didn't see us pulling these sophomoric pranks, and it helped lift morale. This commander did one more thing that became the "secret of Sheppard," one that only the crewdogs who sat alert at Sheppard knew. Why did the same crews keep *sacrificing for other crews* to go to this austere location for a week of boredom? These crews knew that the Det commander had hired a new cook for the alert facility; however, this was not your average cook. This cook was very special and was married to the head of the commissary on base. It was not uncommon for the SAC troops to have lobster or prime steak for dinner during our tour. We swore an oath of secrecy not to divulge this to anyone back at Carswell, and, being the "SAC trained killers" that we were, we almost held our breath each week, wondering when we would be caught and the Det commander fired, since nothing good like this *ever* was supposed to happen to SAC crewdogs. It really boosted morale.

SAC security personnel are, in my opinion, the toughest and most dedicated in all the Air Force. They physically stand guard around our aircraft night and day, regardless of weather, ensuring that *nobody* every breached security regarding SAC aircraft, especially the BUFFs. They were legendary in their firm stance on regulations and enforcing security. We had all heard the story of Gen Cutis LeMay in the early days of SAC testing these security troops by bullying his way by rank through an ECP (Entry Control Point) in his staff car, only to be told to halt by a SAC Security Guard. As I heard it, as he continued to drive his car on through, he was told "Halt, or I'll shoot!" Next, he heard a shot fired as he drove a few feet further, and he was then surrounded by SAC Security troops with police vehicles. Ashen-faced officers arrived to try to remedy the situation, knowing their careers were over for sure, only to find that LeMay was congratulating the SAC troop who had fired at him, saying that the only thing he did wrong was that he did not hit him with the bullet. These guys are *serious*, and they mean business 24 hours per day. Now, add to that fact that our crew was at a somewhat remote satellite facility, and you can see that *nobody* should mess with these guys, not even CINCSAC, himself.

On the first morning of alert, after we finished the assumption of alert briefing, our crew drove out to our B-52, and, after producing the proper challenge/reply code for the day with the security troop, we entered the infamous "No Lone Zone" (complete with a distinctive red-painted line on the ramp around the entire B-52) and conducted our through preflight of all our systems and equipment. We ran into some maintenance problems and ended up being the only crew still out on the ramp one hour later. When we finished, the Nav and Radar Nav were working on something at their station and told the pilot, co-pilot, and gunner to take the truck back to the facility. The rest of us would walk back 100 yards when we were through. As we finished, Dick DeRoos (the Radar Nav) went down the stairs, with John Turner (the Nav) right behind him. I then realized that I left my hat upstairs in the EW station and yelled at John to wait a second. John came back up into the plane as I grabbed by hat and started down the stairs. It was too late! (Note that this all happened within 10 seconds.) "Hey, you guys had better get out here, right *now*!" yelled Dick in a strange muffled sound. John and I emerged from the BUFF, only to find Dick face down on the ramp with a security troop above him with an M-16, and three security vehicles headed towards us with a "blue light special" flashing and sirens wailing. We all ended up face-down on the ramp, and then we all got to talk with the security commander afterwards. Whew! (Dick has not forgiven me to this day.)

About Tuesday of that week, the Det commander gave us a heads up that something was about to happen. The klaxon sounded, we responded to our aircraft, fired up the engines, and found out that this was an ORI (Operational Readiness Inspection) for Carswell AFB. We took our CCP (Command & Control Procedures) tests and completed all checklist items for "un-cocking Alert" status. Another SAC base assumed our alert sorties for Sheppard, and we were told that we would not be going home on Thursday, but would be flying the ORI from Sheppard AFB, then returning to Carswell. Once we recovered from shock and completed all our target study and ORI preparation, we found that we had one evening which we could have a little freedom to drive our alert vehicles to the base BX or whatever. A friend of mine was an IP (Instructor Pilot) on base, so he and his wife invited me to come over for supper. I told the rest of the crew to take the truck without me, and that I was having supper on base. My friend arrived, and I was summoned to the SAC gate to meet my visitors. The wife offered to do my laundry, so I decided to go back into the facility to get my laundry bag. Knowing that my friend knew nothing about SAC security procedures, I had a short discussion with him. I told him about the ECP and the No Lone Zone painted lines, and I warned him not to go near either of them. My IP friend was now a "big shot" 1st Lt. IP on this prestigious UPT base, where he trained both US and German Air Force pilots. He told me not to worry.

I raced into the alert facility to my room, grabbed by laundry bag, and returned not 90 seconds later (after all, I was going to be "free" on Sheppard AFB for one evening!), only to find another of those now familiar "blue light specials!" This time, there were not less than five police cars with lights blazing, and my IP friend was face-down on the ramp with three M-16s pointed directly at him. Eight or nine security troops surrounded the ECP, and even Cutris LeMay would have been in deep trouble at this point. The IP's wife was in tears, yelling, "Gary, help him!"

Apparently, my friend had decided to walk over to check out the ECP and take a look at the BUFFs, despite my warning. He had been challenged to halt and identify himself. Unfortunately, he decided to smart off and refused to cooperate with this enlisted troop. The Det commander, mind you, who was already there (how on earth could this happen so fast?), pulled me off to the side. I explained the situation, and he told me sternly, "Lt. Henley, you had better talk to your friend after you leave here, because he was very close to being in some very serious trouble here. I'll take care of this; now, get him out of here before this gets worse." "Yes, sir!" I replied as I quickly took my now ashen-faced IP friend (who by now was literally shaking all over from fright) and quickly departed the SAC alert facility area and returned to his house. He tried to pretend that he was OK, but it was evident that he had experienced a "significant emotional event." Other than a very careful return to the alert facility later that night (where my friend now listened carefully to every word of my instructions as we approached the SAC area), I don't expect that my cocky 1st Lt. IP friend ever went near a SAC alert facility or BUFF again in his entire life. He gained not only a very healthy respect for the seriousness with which SAC protects its assets, but also a unique appreciation for the operational Air Force beyond ATC. Life in BUFFs is a whole lot different than life in T-37s!

The author in the BUFF's pilot's seat.

And It Wasn't Even an Eskimo Pie

George Golding

Late one afternoon while preflighting our B-52F in the pouring rain for a Chrome Dome mission, we made an error in loading the flight lunch chests. Since this mission was planned to be airborne in excess of 24 hours, the inflight kitchen had issued us two chests of food for the front crew compartment and one for the gunner's compartment located in the rear of the aircraft

In the confusion of trying to complete the loading of our flight gear, including the ice chests, we made a slight mistake. As we leveled the plane off at 29,000 feet heading for our orbit track, the gunner got hungry and opened his ice chest to get some food.

Over the intercom came a stream of words that were not all that professional. My CP asked the gunner what was wrong. It turns out that one of the ice chests that was destined for the front cockpit was filled with only ice cream cups. You guessed it, we had the gunner's real food and all he had was our ice chest full of ice cream. This did not make him happy.

Due to the restrictions involved in this type of nuclear airborne flight the two person rule applied. That meant that for us to be able to swap the ice chests required that the EW and the Navigator to have to crawl on a tiny ledge through the bomb bay to the rear of the aircraft while dragging the “food” ice chest and then return with the “ice cream” chest. A maneuver like that required too much time and effort. Such a plan would also require making a descent to 10,000 feet while this happened. Our B-52 just didn't have the extra fuel required for the unscheduled descent and the climb back to our designated orbit altitude.

As pilot, I had the unfortunate duty to have to tell the gunner he would just have to eat the ice-cream for the rest of the flight. The thought of having nothing to eat but ice cream while orbiting in circles near the freezing North Pole seemed extremely comical in a way.

For the next 26 hours whenever the gunner called in his oxygen check, he would say, "Ice Cream...I've got Ice Cream...Lots of Ice cream!" On future missions of that type, the gunner never again allowed the copilot or anyone else on the crew touch his ice chest.

Rookie Alert Tour - SAC ORI on Final

Lynn Wakefield

1973: HQ SAC decides to implement a policy of forming integrated crews and training them as a crew at Combat Crew Training School (CCTS), then send them to their first duty station as an integrated crew to fly, certify, and perform as a line crew. The rationale was that if we trained from day one as a crew, we would pay dividends earlier as a better, more coordinated crew versus training individual crew members then building crews and crew coordination at duty locations (down stream in the aircrew training life cycle).

R-21 was one of the first integrated B-52G crews to train as a crew, and graduate as a crew from CCTS, Castle AFB, CA, in December 1974. We reported as a complete crew to the 42d Bomb Wing, Heavy, Loring AFB, Maine, in January, 1975.

We arrived, flew local orientation flights to prove that we really could fly the G-model, certified our Emergency War Order (EWO) sortie to the Vice Wing Commander, and joined the line as crew R-21 - there were only 21 numbered crews at Loring so we knew where we were in the seniority pool. Hey, we didn't care if that meant that our crew leave came in October and April, we were line crew dogs, and not in training flight. In SAC, it is important to find the guys behind or below you - that way you know where you are in the formation.

We went our first alert tour on a Thursday in Mid-April, 1974. On that following Monday, the Aircraft Commander, Capt Tim Lyle and I are sitting in the cockpit while maintenance is working on something. We watch a KC-135 land, however, it is unusual in that it has no boom to transfer fuel. Two rookies have just witnessed the arrival of the HQ SAC Inspector General Team - they are here to evaluate the Wings ability to fulfill its wartime mission responsibilities - an Operational Readiness Inspection (ORI) is about to begin.

As crews on alert, we take Command and Control Procedures (CCP) tests, Emergency Procedures (EP) tests, have our EWO (wartime gear on the aircraft) bags checked for compliance and completeness, and then fly the ORI sortie, which is a overland US sortie designed to simulate an actual sortie that could be flown in wartime.

The sortie profile included air refueling, low-level navigation, precision bomb delivery, electronic warfare defense exercises, enroute navigation, and radio discipline.

As R-21 took off as scheduled to fly the ORI sortie we were feeling good having tested well. Our gear passed inspection - we rookies were looking good across the board - confidence was high. All good until we rotated and the co-pilot said, "gear up - uh, oh."
The right front main gear failed to raise up and lock. We climbed to an intermediate altitude and ran the checklist - no joy - the gear would not raise. Under the rules of the day, if we called back for guidance or help, we would have violated CCP rules of radio silence as we were simulating being "executed" - any breakdown in communication discipline could result in our sortie being "terminated" and all of our simulated bombs and other scored activities on this flight would be counted as a zero in the wing's overall score - not good on any grading scale.

So we elected to lower the gear into the down and locked position and then pull the circuit breaker to keep it (hopefully) in that locked position. As we could not keep up with the rest of the string of Loring bombers (our airspeed was limited due to the hanging gear), we sucked as much fuel from the tanker as we could, flew at a lower altitude than the rest of the string and entered low level as the last bomber in the flight string.

The Nav and RN were working their tails off downstairs doing some ad hoc mission planning, the Co-pilot was eyeballing the fuel management system and calculating the new, elevated consumption rate versus our ability to get to our targets, the pilot was talking to air traffic control and trying to keep every airliner on the east coast from making us a tour stop for their passengers (e.g. "Off our right wing, you will see a B-52 with its gear hanging, not the normal flight configuration for the world's most powerful weapons delivery platform…"). My job was to call someone, anyone on HF, to advise them as to our problem, what our solution was, and what our plan of attack was going to be. Our rationale was that if the EW made the call, no one would order him to land as he was not the Aircraft Commander, and the signal on HF is famous for fading in and out, so I might not hear any recall words. We did not want to get pulled out of the fight early - we were rookies, but we were not dumb - no wing could afford to have all those lost bombs - everything counted.

We fly the route, complete all facets of the sortie, except for the enroute celestial navigation leg, declare an in-flight emergency (IFE), and recover at Blytheville AFB Arkansas. When we land, there is no welcoming committee - no one wants to talk to us. After the maintenance debrief we spend the night in billeting, file a flight plan direct to Loring AFB, climb in the jet and launch for home. Maintenance said that all continuity checks had checked good, the gear should work just fine. Sure enough, when we raised the four main landing gear, three of them responded as advertised - the right front remained in the down position. This time I found the circuit breaker immediately and pulled it with practiced authority. We flew the nav leg and recovered at Loring AFB around sundown.

This time, we had a large reception committee. Alpha (Wing CC), Bravo (Wing CV), Charlie (Wing DO), and Delta (Wing Deputy Commander for Maintenance) were there along with about 100 chiefs and assorted maintenance folks. Apparently, there was going to be some serious trouble shooting about to begin. A B-52 weighs about 200,000 pounds empty - the jacks required to raise it are sizeable - you just know when they are around.

We debriefed in mass as is usual practice at maintenance, then we debriefed in mass for the squadron staff, which included every division chief (Bomb-Nav, Defensive Counter measures, Training, Stan/Eval) as well as the squadron CC and DO (not usual).

Then we debriefed by crew position individually with our specialty division. (It turns out that the HF call that I made was not to a base Command post at all - I had raised LOOKING GLASS on the radio who passed all to the Loring CP. Well, he did say that he had "Extensive relay capability.") The IG had a representative at every meeting. If there was a hole in this story someone was going to find it.

At the end of the next day, we were summoned to the Squadron CC's office. Our Commander told us the following: The IG took our information and flight data and re-plotted everything from the ORI sortie against our real EWO sortie using the worst case winds and worst case fuel consumption/degradation factors due to the hanging right front gear. The master navigators on the SAC IG team decided that we would have dropped our last real bomb with about 4,000 pounds of fuel to spare - just enough climb up and maybe bailout, maybe. But as long as the bombs got out, our sortie was considered good and all scores counted - all our ORI bombs, nav legs, ECM runs, and test scores were good. So R-21 went from rookies, to the edge of the professional toilet, to the top of the instructor upgrade lists all in about six months.

Aerial view of the abandoned Blytheville AFB, Arkansas, alert parking Christmas tree and alert facility (square area at mouth of parking area) as it looks in 2005. This was a typical Alert area layout on most SAC bases.

Bite Me Crank, Mate!

Tommy Towery

During our morning alert preflight one hot day in Texas, my crew and I heard over the command post radio that an unannounced KC-135 was on final. Since we knew our wing was imminent for an ORI, we watched with interest as the plane landed and proceeded to taxi up to the front of base ops. When the engines were shut down, a team of blue suited officers began piling out of the newly arrived plane. I had been preflighting the sextant and put the crosshairs on the team leader as he immerged like a sniper would have centered his sight on his target. Had it been a sniper's rifle, I could have nailed him. That's just how happy I was to be on alert when an ORI hit. A minute later we saw our wing commander and his staff show up and greet the visitors and the congregation proceeded to shake hands and share the most common lies swapped during any inspection visit in SAC – "We're here to help you" answered by "We're glad to have you."

Knowing that the klaxon would go off in a few short breaths, I and my crew elected to drag out our preflight longer than necessary and stay aboard our alert aircraft until the horn sounded. Our wait was not long, and we were sitting ready to decode the emergency action message as soon as it was broadcast. Needless to say, we were the first crew in the wing to report cocked and ready.

We were all instructed to wait aboard the plane for the inspectors to arrive. As we waited we decided to quickly inventory our equipment and be ahead of the game when they arrived. My Radar Navigator at the time was Bill Tooker. Bill was a great guy to be around, and shared my sense of humor – which is a bit odd. Bill's dad was an active duty USAF general, but that was not normally known to folks. To everyone in the wing he was just a normal Crewdog like the rest of us.

So, there we were, waiting on the inspectors to arrive and getting really bored. I don't know why, but Bill was looking through the EWO gear and found the pilot's eye patch which was designed to protect at least one of the pilot's eyes from flash blindness in case of a nuclear detonation. Just as I was finishing checking out my EW crew position, the quietness of the cockpit was interrupted with the sounds of singing. "Yo-ho, yo-ho. A pirate's life for me...." I crawled out of my station and looked around the corner to see Bill, sitting on the sextant case with his flight cap on sideways and the eye patch over his eye playing an air accordion. He was rocking back and forth and continued with the pirate song from *Peter Pan.*

Bill was off headset in his merriment and was not aware of what else was going on in the aircraft. He turned around and saw me and blurted out in a loud blustery pirate accented voice, "Bite me crank, mate!" just as one of the IG officers stuck his head up through the hatch.

I sat petrified, awaiting the wrath of God, or at least his right hand man – the Inspector General staff officer – to strike me dead. He angrily looked at me and was a little puzzled as to why I would say such a thing to him. I froze. Hearing the "Yo ho, yo ho" singing start up again he quickly realized that I was not the one from which the comment came and turned to see Captain Bill sitting there, eye patch, crooked hat, and all still playing his air accordion and singing.

It was that moment that I was made aware that even IGs were once Crewdogs themselves. He looked again at Bill, then looked back at me, kind of shook his head and burst into laughter. Lucky for us, we had all our ship in order, even if it was a "pirate" ship and nothing was ever mentioned in the inspection write up about the unprofessional crew incident aboard a B-52 bomber on nuclear alert. But there were sure some bar stories told about the incident for years to come.

Known-Notice Alert Exercise

Bill Beavers

The setting: 5th Bomb Wing, B-52Hs, Minot AFB, ND in mid-1970s; at the "Alert Pad/Bomber Alert Facility" about 7:00 P.M. one December Sunday evening.

It was cold and dark and windy outside (maybe somewhere around zero degrees Fahrenheit) and the Buff Crewdogs on alert were mostly finished with dinner and most of us were sitting around in the Alert Facility in the TV Lounge.

We all know that the Alert Force was to be kept unaware of when the Klaxon would blow for an alert exercise and we would all scramble to our planes and start engines. But the senior Wing Staff, among others, was advised according to a weekly schedule published by HQ SAC of what time and date each specific Base's Alert Force would receive a message that would cause the local Command Post to sound the Klaxon and thereby cause us to scramble to our planes. The senior staff had to know so that they could be pre-positioned to observe and monitor our scramble at the time the Klaxon went off. But we Crewdogs "were never told ahead of time" for that would spoil the Exercise results, particularly the time it took from Klaxon sound to specific events (first and last airplane to get engines started, "Ready to Taxi" times, and if it was a "Mover/'Coco'" exercise first and last aircraft across the runway Hold Line, etc.). So the Crewdogs and the Senior Wing staff always had a running contest of the Crewdogs being able to find out when the next Klaxon horn would sound, ahead of time, and the staff denying us that information (although, for years, it was common practice to say that our wives knew the Exercise times before anybody else did).

Back to my setting at the 5th Bomb Wing Bomber Alert Facility at Minot that Sunday evening. The Crewdogs managed to "Turn the Tables" on the Wing Commander, Deputy Commander for Operations (DCO), Deputy Commander for Maintenance (DCM) and other assorted Colonels and Senior NCOs that evening. We knew exactly when the Klaxon was going to sound for that night's Alert Force Exercise, down to the minute. And we knew about two hours ahead of time. I will not divulge how we found out, just suffice it to say that the

Crewdog "Intelligence Gathering Network" was far better than the senior staff's COMSEC (Communications Security).

Anyway, our Colonels showed up that cold Sunday evening about 10-15 minutes ahead of scheduled Klaxon sounding time. Some of them stayed in their warm cars and parked strategically around the Bomber/Tanker Alert Pad to "observe" the pending "No-Notice" Alert Force Exercise. But the Wing Commander and DCM personally came casually strolling into the Bomber Alert Facility a few minutes before Klaxon time. Now why would both of these Wing senior officers be out at the Alert Facility on a Sunday night just "casually strolling around?" Didn't matter, we already knew why. And until they strolled into the TV Lounge where the Crewdogs had all gathered. They didn't know we knew why they were there on a Sunday evening.

There we all sat, stood, walked, leaned inside of our nice warm Alert Facility "casually watching TV" --- all of us with our Parkas, Mukluks, and gloves on over our Nomex flight suits !!! "Gee Sirs, what brings you out here on a cold Sunday evening? Us? Well, yes sir, we all wanted to be in the proper uniform when we heard you were coming."

The Klaxon DID sound in a few minutes and we "did our thing" and they "did theirs" and both sides had a good laugh about it afterward.

20th BS Carswell R-04 (L-R) Denny Scruggs, Tom Adcox, Dave Kerr, Cloyce McClean, Tommy Towery, Doug Bunch

Santa Don't Need No Stinking Line Badge

Denny Scruggs

One Christmas Day when my crew was on alert, we had one of those rare occasions when the families of Crewdogs were allowed to come through the security gate and visit us in the facility. The crews were still "on alert" and if the klaxon sounded the rules were still the same about getting to the aircraft for immediate launch. We did feel fairly safe that we would not have an alert "exercise" during that period but we had to be vigilant awaiting the real call to go to war. I seem to remember that the families were also allowed to eat in the alert facility on that holiday as well, and taste the type of food we put into our mouths about 120 days a year. Our Wing Commander, a stern officer with little or no Christmas spirit, escorted the gaggle into the alert area. Dave Love, a fellow pilot, and I were playing pool in the recreation room when the guests arrived. Our Colonel walked over and we politely offered a cheery "Merry Christmas, Sir!" He replied, "Scruggs, you and Love be careful, that is brand new felt on that pool table, I do not want it torn!" So much for Merry Christmas, but the fun was just starting.

Bachelor Crewdogs found better things to do than to sit around while wives and kids invaded "their" space, but for the families, it was a way to share Christmas day in a very special, but still strange, way.

This Christmas visit sticks in my mind for one very special reason. Many of the visiting families brought out the Christmas presents from under the tree at home to share the opening experience with the Crewdogs. My Radar Navigator at the time, Dave, was married to a woman who was, by her own admission shall we say - different. She was a wonderful person but sometimes did unique and unexpected things. When she arrived at the facility she had brought one present for each member of the family to open, including one for her husband. There were several small boxes and one long one. She and the kids and Dave went down to one of the semi-private mission planning room to open the few token presents.

A little while later, Dave came around looking for me. Dave had a problem. Tommy Towery, my EWO, and I went with him into the small room and there lying on the table was Dave's brand new Christmas present - a Remington pump shotgun. Of all the presents to bring into a secure military facility, she had elected to bring him the 12-gauge shotgun he had been wanting, fully assembled, ready to lock and load. There were many things Crewdogs were not allowed to possess in the alert facility such as alcohol, drugs, and of course, weapons!

The way he had opened the present had shredded the long strips of Christmas paper and with the Remington Logo and picture of the shotgun on the cardboard box, he had no way to conceal the weapon to get it back out of the facility and out of the secure area. Dave and I saw our military careers going down like the captain of the Titanic. As a SAC Aircraft Commander, I was ultimately responsible for everything that happened to any member of my crew.

We thought a while and finally decided to break it down and place parts of the shotgun into the sleeves of our jackets and carried it out to the alert truck and just drive it out the gate. Whew! That event was almost as bad as the time my copilot, Tom Adcox, brought a hunting bow and arrows on alert with him at Wichita and was caught on the perimeter at night shooting at Jack Rabbits. Oh, my aching back. It was right up there with the time that that Tommy and the navigator, Cloyce McLean, brought a portable lead smelter to the pad and made bullets for their muzzle loaders down in their room at night. I knew I smelled something burning!! Or the time Tommy took all the golf balls in the alert recreation room and stuffed them into the navigator's pillow starting an EWO versus Navigator war of no small proportions that went on for months. You have to remember that being on Alert was very, very confining. Can't we all just get along? I digress too much.

It was indeed a Christmas to remember. Probably best of all, the present to the rest of the crew was a great story to remember for years. I wonder what the statue of limitations is for unauthorized weapons on the Alert pad.

When a Good Message Goes Bad, Very Bad

Lynn Wakefield

In The Spring of 1980, two incidents occurred that rattled SAC from top to bottom: In one instance a computer chip failed in the Cheyenne Mountain Complex which triggered the Alert fleet to start engines. In the other incident, a "Practice" alert message was somehow sent as a "Real" message and the fun began. My crew, S-01, was sitting a GIANT LANCE (GL) Alert sortie at Carswell AFB when this second incident occurred.

Some sorties did not execute on launch. They went somewhere else, waited, and then did what ever needed to be done after the first wave went through. This roughly explains what a GL sortie did.

When the klaxon went off, we charged out to the aircraft as always. It's always real until it's not—SAC's golden rule. On our crew, the decoding responsibilities fell to the Radar Navigator, a California guy with four children, and me, the Electronic Warfare Officer. Pilot and Co-pilot were waking up the bird in preparation for engine start; the Navigator was pulling four engine covers while the gunner was pulling the other four engine covers and rear chocks. The crew chief had the ground generator, front chocks, and the pitot covers.

NOTE: Even after 25 years, as I write this I can clearly remember how orchestrated our actions seemed—we didn't even talk about it—we just executed the ballet.

As the RN and I unlock our respective locks (2) on "The Box, " I copy the message as he grabs the decode book. But, this time we need the "other set" of books—the "Real" set, the "It's not practice" set of response books.

His color noticeably changes, but we say nothing—open the books, "pop" the correct tickets, decode the message, and read the response instructions to the crew. We do it again, and then a third time. The pilots are firing the engines as the instructions take us to the hold line, that area just before turning onto the active runway—the place where the Airline pilot tells the flight attendants to "Take a seat and…."

As we are in the throat of the alert pad, we taxi first and are at the hold line first, ready to go. Here we sit for over two hours waiting for the next message which will launch us loaded into the night to do whatever needs doing—it's what we are trained for. But now we have a little time consider, review, and replay.

In the first hour, I estimate that the RN and somebody decoded the message at least 19 times—he just could not believe that he was going to have to do what he had trained to do. While he was thinking over this issue, the Co-pilot, 1st Lt "Ramjet" Riordan and I also had some thinking to do. As we both smoked at the time (hey, it was 1980 after all), we inventoried our cigarette supplies. We had about two packs between us total. By my calculation, one of us was going to have to die just after air refueling if either of us was going to survive the flight at any speed.

After a couple of hours we got a recall message (as we were not launched, we did not have to be terminated—which would further rattle Offutt—remember your CCP, you know what I mean here). After the proper decoding and authentication, we accomplished a "baby elephant walk" down the runway and drove back into the alert pad. It took about 6 hours for all recovery actions to complete as 9 bombers and 4 tankers had to refuel, and check brake and hydraulic fluids, find our six-pax trucks, and, of course, debrief.

Our RN was never exactly the same again. He and his wife had one more child and he resigned his commission a year later. Ramjet and I made a beeline for the BX the next morning and each bought two cartons of cigarettes for our EWO bags. Bad message or not, we had learned our lesson—it's one thing to launch the fleet, it's another to do it cold turkey!

Responding to the klaxon.

Gentlemen, Start Your Engines

Denny Scruggs

One lasting war story about my alert days will forever stick in my mind and this is no s---. One time, during one of my many alert tours, I was taking a training ride with my Squadron Commander in the Carswell B-52 pilot simulator which was located away from the alert facility. Naturally, I was really concentrating on doing all the emergency procedures by the book, when the SAC command post broadcast a radio message for the alert crews to report to their aircraft. It was a message that I completely missed, the squadron commander missed, and the guys running the simulator ignored. Somehow they seemed to forget that I was on alert, and although they heard the message, they did not tell me.

About 10 minutes had gone by after the missed message, when then the next thing I hear is the infamous "Klaxon" horn resounding through the building and the famous words "Klaxon, Klaxon, Klaxon" broadcast on the radio. Those hallowed words were followed by a series of alpha numeric letters that told you basically, get to the plane and start engines NOW!

OK, by that time the adrenalin was really pumping. I run to the Maintenance van I was using, since our crew alert truck with the rotating beacons and siren was broken or out of service or somebody on my crew had it besides me. Whatever the case, I was using a non-standard truck as my form of SAC Alert transportation. I squealed rubber and burned out of the parking lot only to be immediately pursued by our attentive security police with their own sirens wailing. Do I stop? Heck no! The chase continues to the alert pad where I flash the secret signs to the armed alert guards and they waved me into the parking area. "This will be easy, my plane will be the only one with engines not running", I say to myself.

Now you have to remember that Carswell was the biggest bomb wing in the Strategic Air Command during that period. Where a normal base would have six bombers on Alert, Carswell would have up to 12, and they are all camouflaged with dark red numbers painted on big black tall tails. I drive the line looking for a B-52 waiting to start engines. To my bewilderment, all the aircraft have engines running and hatches closed. "This is not going to be a good day", I say to myself as I drive the van behind the blast fences, wondering if perhaps my presence on the alert area had been noticed... right! Eventually I spy my aircraft and lo to my amazement I look up and there I sit at the pilot's window in my uniquely painted flying helmet bobbing up and down. It is being worn by someone, could it be me, giving a "thumbs up" out the pilots side window to our "Your only as good as your last mistake" Wing Commander of the 7th Bomb Wing Heavy.

At that moment I had an out of body experience and viewed the rest of this disaster looking down from above.

I saw myself stumble from the van, the crew chief and security guard clear me into the entry hatch, I fumble with the latch, the hatch opens, I scramble inside, and no one will look at me. I climb the ladder to the upper deck and there sits TT, my EWO, manning the pilot seat along side my copilot Tom Adcox.

Although, to this day I have no doubt that Tommy Towery could have flown that airplane (he did own and fly his own plane), the Strategic Air Command had some pretty well defined guidance concerning who could and who could not occupy the B-52 pilot's seat during SAC alert engine starts. EWO's were not on that privileged list.

It seems that in the confusion of having an alert engine-start exercise without the pilot bothering to show up, my crew had a little democratic discussion. Since the copilot was the one who, according to the checklist, actually started the engines, there was some contradicting confusion. Should they sit there and do nothing while everyone else prepared to go to war or at least have the plane ready to go when the pilot arrived? After a crew vote, the decision was made. Start the engines!

Well, the crew got an A+ for initiative and nice try, however, I got a big "Aw Shucks" and a lot more face time with the wing commander than I really wanted or needed to get. Plus I got an opportunity to train SAC aircrews on just exactly who has to be in what seat for an SAC Alert Aircraft Start up.

I never missed another alert, ever, ever, ever.

The Day We Thought the Klaxon Sounded "For Real"

Tommy Towery

I was lying in my bunk in the Alert area, assured by the Wing Staff that we would not have a practice exercise. "If you hear the horn, it's the real thing--we're going to nuclear war."

During the whole period we sat nuclear alert in the B-52, we never really expected to hear the horn that would send us off to a real atomic war. That was one of the things that allowed us to get through each day of alert, even though that mission was what we were trained for and prepared to do. We always knew when the klaxon sounded that it would be an alert exercise. The scariest option to that scenario was that it would be a no-notice ORI exercise starting and that we would be tested and inspected for the next few days.

That was the way I felt almost every time that I inventoried a mission bag and my crew assumed the duty. That all changed one day and the sound of the horn sent chills through my body and made me prove that I was ready for the job.

It happened on a deployment to Guam after the bombing had been halted in Vietnam. Crews were still being rotated from stateside bases to Andersen AFB and U-Tapao. Since the 43rd Bomb Wing had not been brought back up to speed and fully manned and ready to assume their own nuclear alert, that task fell to the visiting TDY crews.

I remember that some political issues came up while R-04 was there on our three-month deployment. As soon as we assumed alert, it was announced that SAC had sent down an order that put the entire worldwide alert force in an increased alert readiness state. The 43rd Bomb Wing staff called all the crews into the briefing room and informed us that during this period all practice exercises were going to be curtailed. They made it clear that we would only hear the klaxon go off if there was a real nuclear threat.

Even though we were a young crew, we knew that something serious was going on and that this order was neither to be questioned nor taken lightly. It was an uneasy feeling among the crews throughout the BOQ (Bachelor Officers' Quarters) that served as the temporary alert facility. The next day we were startled into reality by the sound of the klaxon, and our heads filled with thoughts of "This is it!" We were deathly quiet as we manned the trucks and rushed to the aircraft to start engines. As soon as we got the aircraft powered up and got on headset, the command post quickly came over the radios and cancelled the alert and ordered the pilots into the briefing room.

The rest of the crew remained curious as to what was going on until our pilot came back with the answer. It was explained to him that a tropical rainstorm had allowed water to seep into the klaxon control box setting off the klaxon unintentionally. Apparently it was not unusual because of all the rain that Guam received and that this malfunction had happened many times before.

They were going to do something to keep it from happening again. The pilots were told that the klaxon response system was going to have all the power to it turned off, and it would be turned back on only in case of a real nuclear war. We went back to our alert quarters a little shaken up from the drill, but feeling a little easier, after being assured that it would not happen again, unless we were going to launch on a real nuclear mission. The political tension was still high, and we all knew it.

So, there I lay, assured that we would have no "exercise". It was guaranteed. That's when we were again jarred into the reality of a possible war when the klaxon horn sounded again. The sky was blue and clear, and it was not raining. We remembered the words of the commander, "The horn will not sound unless we are really going to war." This time we were not quiet, and we discussed all the options as we scrambled to our cocked aircraft. We rushed to the planes, thinking we were really going to launch for war and were headed for our last B-52 flight.

This time, we were not stopped from starting engines (that was bad). "That's not a good sign," I remember thinking. We started taxing to the hold line. "Crap, this is it!" we thought. As soon as all the BUFFs were out of the alert area and headed for the hold line, once again the command post announced over the radio that the launch was cancelled.

We were told to taxi to a specific area on the ramp and to hold there, to stay on radio, and to await further instructions. Time passed, and finally we were ordered to taxi the aircraft back to our parking spot.

In another briefing that followed, the pilots were informed that a transient C-141cargo aircraft had been parked right behind the alert aircraft and had developed a ground fire in its auxiliary power unit. The command post was afraid that the C-141 might explode/catch fire and that the alert aircraft were in danger of catching fire as well. Their quickest solution in this situation was to blow the klaxon to get the alert crews to quickly respond to the area and move the alert aircraft out of harm's way.

Crew tensions remained, and the klaxon did not blow at all the rest of the tour; however, we were on edge the whole time. It was the closest we ever came to going to "the big one," and I will always remember the time we heard the horn to go to nuclear war.

Combat in the B-52 –

The War in Vietnam

Vietnam and Linebacker II
Charles N. Brown

The US Air Force first became involved in Vietnam in 1950. B-52s began flying missions to South Vietnam out of Guam on 18 June 1965, a twelve-hour flight. These were F models converted to carry fifty-one 750 lb bombs (27 inside and 24 outside). The usual white nuclear anti-blast color underside was painted black. This made them less visible from the ground. In those days the top half of the plane was silver with the standard SAC and USAF stars and stripes markings. Later all B-52s were painted camouflage colors on the tops with the Vietnam iron bombers given black bottoms and the nuclear carriers white. The current painting is solid dark. In 1966, older D models replaced the Fs at Guam. These older Ds were given a "big belly" modification that allowed them to carry 108 (84 inside and 24 outside) of the 500 lb and 750 lb bombs.

In two years, B-52s flew 10,000 sorties to Vietnam from Guam. In 1967, B-52s began flying out of Thailand. Missions were only three to five hours versus twelve hours from Guam, and heavier bomb loads could be carried since less fuel was needed. Workload on crews and maintenance were less. In-flight refueling wasn't needed regularly. These daily schedules continued for another five years. The steady increase in US military involvement in Vietnam caused considerable unrest among people at home, and years of peace negotiations were ineffective.

During the last half of 1972, there had been a buildup of B-52s on Guam to augment the B-52D models in Thailand. Whole wings of US based nuclear bombs, support people, and support equipment were deployed to Guam, including those from the author's 97th Bomb Wing at Blytheville AFB, Arkansas. By December there were fifty-four B-52Ds in Thailand and 152 B-52s (53 D models and 99 G models) on Andersen Field in Guam for a total of 206 bombers avail-able in the theater. In addition, there were now 450 tactical support aircraft assembled in Thailand: F-4s, F-105s, F-111s, A- 7s and EB-66s.
In the fall of 1972, negotiations between the US and North Vietnam to end the war were being held in Paris. On October 8th, Henry Kissinger indicated finally that North Vietnam accepted virtually all American terms including a cease-fire and return of all US POWs. On October 18, Kissinger flew to South Vietnam for coordination. On October 24th, he announced in Washington, "Peace is at hand." On November 20th, both sides returned to Paris, but North Vietnam appeared to stall.

Finally, on December 4th, the final session in Paris, the North Vietnamese rejected previous agreements on key issues, and negotiations fell apart. President Nixon then told Kissinger in the oval office, "If we renew bombing, it will have to be something new, and that means we will have to hit Hanoi and Haiphong with B-52s. Anything less will only make the enemy contemptuous." On December 15th, the Joint Chiefs of Staff told the Pacific Command to prepare for maximum effort strikes. This action at Guam and Thailand meant no more B-52 take-offs on routine missions. All aircraft were recovered, which took up to twelve hours, and they were loaded and prepared for a new mission assignment. On December 16th, in Washington, Kissinger announced the talks failed to produce an agreement. The following message was sent from the Joint Chiefs of Staff to Pacific Command on December 17th: "You are directed to commence at 1200Z, 18 Dec 1972, a three-day maximum effort, repeat maximum effort, of B-52 and tactical aircraft strikes in the Hanoi-Haiphong area on the authorized target list. Be prepared to extend operations past three days if directed."

The "Linebacker II" plan was basically daylight low-level tactical aircraft strikes into North Vietnam coupled with high altitude B-52 strikes at night. Targets around Hanoi and Haiphong and its harbor were railroad yards, storage areas, radio communication facilities, power facilities, airfields, SAM (surface to air missile) sites and bridges. North Vietnam possessed the best air defense system in the world, created from Soviet equipment. They had excellent radar, surface-to-air missiles with 200 launchers, 4,000 anti-aircraft artillery guns and 200 MiG fighters.

On the first night, December 18th, 129 B-52s attacked military targets in three waves about four hours apart to provide the psychological impact of continuous bombardment. Each wave had 40-50 bombers. The B-52 flew in three-ship cells with 500 ft vertical and a one-mile horizontal separation between aircraft. Two to three minutes separated the three ship-cells. The first casualty of Linebacker II was a B-52G from Blytheville, Arkansas, (the author's home base) in wave one. At the bomb release point two SAMs hit it simultaneously. The aircraft nosed down, crashed and exploded ten miles northwest of Hanoi. The pilot, Lt Col Rissi, and gunner, MSgt Ferguson, lost their lives due to the SAM detonation. The copilot, 1Lt Thomas, was missing in action (MIA). Radar navigator, Maj Johnson; navigator, Capt Certain; and electronic officer, Capt Simpson, successfully ejected, were captured, and became SAC's first POWs. Their pictures were publicized by the North Vietnamese and were reprinted in both Time and Newsweek a few weeks later. They were released when POWs were returned. Through the course of the next ten nights of combat, fourteen other crews shared this crew's fate in varying degrees. The first night's 129 sorties ended with three B-52s downed and two badly damaged. This was a two and a half per cent loss rate. SAC planners had expected higher losses. Over 200 SAMs were fired.

The second night, 19 December, consisted of 93 bombers. None were shot down in the 180 SAM firings. Two were damaged but limped home OK. On night three, 20 December, it was a different story. Four B-52Gs and two B-52Ds were downed with one more "D" damaged. Again, over 200 SAMs were fired. One of the aircraft shot down was the lead B-52G in wave three. It was hit by a SAM while in its high bank turn after bomb release. Three of seven on board ejected successfully: pilot, navigator, and wave commander, Lt Col Keith Heggen, the author's roommate. The first two were returned after the war as POWs. Lt Col Heggen died in captivity after ten days from injures due to the SAM explosion.

Two cells behind Heggen's aircraft was a B-52G from Blytheville AFB which was hit by a SAM in the forward fuselage. The pilot, Capt Craddock, struggled with the aircraft after a brief dive and was able to regain control when a second SAM hit it. Bailout was ordered, but the

aircraft exploded. The gunner, SSgt Loller, was the only one to eject successfully. He was taken prisoner. The other five were MIAs.
Things improved the next three nights with changes in tactics for support and B-52 aircraft, and reduction in North Vietnam capabilities. No B-52s were shot down those three nights, and only one was damaged.

At Christmas time, President Nixon called a 36-hour cease-fire as a goodwill gesture and observance of Christmas Day. Aircrews needed the rest, and maintenance needed the extra time to fix the aircraft. The December 26th offensive plan was different. One hundred twenty B-52s were massed into a single wave formation to strike ten target complexes from different directions and altitudes at the same time. This was the largest formation of B-52s ever, and nothing like it had been done since WWII. The bombing plan was done by computer at SAC headquarters, Offutt AFB, Nebraska. B-52s dropped 9,932 bombs that night. Two aircraft were shot down by 68 SAMs.

KC-135 tanker support for December 26 was noteworthy. There were 194 KC-135s in the theatre. In 24 hours they made 763 refuelings. The majority, or 607, were in support of tactical aircraft.
The 9th, 10th, and 11th nights saw lighter assaults. Only one aircraft was lost on the ninth night, December 27th. The last B-52 of Linebacker II landed at Guam shortly after noon, 30 December 1972. In those twelve days, with eleven nights of bombings, we lost fifteen B-52s from over 1,000 SAM firings and AAA. Thirteen tactical aircraft were also lost. The total number of SAC B-52 crewmembers involved in the shoot-downs was 92. Fifty-nine (two-thirds) were recovered right away or as POWs. The rest were killed or were MIA. Other non-B-52 aircrews shot down totaled 29 with 17 recovered and the remainder killed or MIA.

What was the political result? A British expert on Asia, Robert Thompson said, “In my view, by 30 December 1972, after eleven days of those B-52 attacks on Hanoi, you had won the war, and it was over. They fired 1,242 SAMs and had none left. They and their rear base at that point were at your mercy. They would have taken any terms. And that is why, of course, you actually got a peace agreement in January which you had not been able to get in October.”

One member of the US delegation to the Paris peace talks said: “Prior to Linebacker II the North Vietnamese were intransigent, buying time and refusing to discuss even a formal meeting schedule. After Linebacker II they were shaken, demoralized and anxious to talk about anything." Henry Kissinger said, "There was a deadlock in the middle of December, and there was rapid movement when negotiations resumed on 8 January." These facts have to be analyzed by each person for himself. The talks swiftly produced an agreement that ended the role of the US in Vietnam and brought the POWs home as well as most of the deployed aircraft and people. But of course, there was a price, not only in lost aircraft and tax dollars, but also in lost crewmembers who are still remembered by their fellow fliers, their families, and friends.

A Picture Is Worth 444 Words

Bill Beavers

The photo above is a picture of Carswell E-57 and their B-52D. It was made at U-Tapao RTNAB (Royal Thai Naval Air Base), Thailand, just prior to a mission to bomb Hanoi on 26 December 1972 during Linebacker II. Left to Right: SSgt Ron Ragan - Tail Gunner; Capt. Bill Beavers – Navigator; Maj. Arthur Craig Mizner - Aircraft Commander; Capt. Bud Hughes - Electronic Warfare Officer; Capt. Don Craig (with headset on) – Copilot; Capt. Chuck Bristol - Radar Navigator.

The bombs are 500 lb. HE (High Explosive) mounted on a MER (Multiple Ejector Rack). The MER rack itself was the beam (with the number "034" on it) you see ATTACHED to the bottom of a AGM-28 "Hound Dog" pylon which was attached to the underside of the wing. There were twelve bombs per MER rack with one MER under each inboard wing section. The release of the (24) bombs from the MER racks was controlled by a SEPARATE BRIC (Bomb Release Interval Control) from the "normal" INTERNAL BRIC (that controlled the rate of release of the internal bomb bay weapons).

The red streamers hanging from the bombs were attached to the bomb safing pins that at that moment were still installed. The safing pins mechanically precluded the bomb release lugs from tripping, opening, and releasing the bombs (one pin per bomb shackle). If one looks EXTREMELY closely one can see the bomb arming wires protruding along the top surface of each bomb and extending out over the bomb nose and going THROUGH the little arming fuse propeller. These wires (one per bomb) precluded the arming propeller from turning and thus arming the nose fuse. In flight and over the target (hopefully) these wires were RETAINED physically on the MER rack and the bombs fell away. As the bombs physically departed the aircraft the arming wire was withdrawn from propellers, allowing them to "wind vane" turning several times on their descent and thus arming the nose fuse. Thus, when the bomb hit the ground it was already armed and would detonate. Sometimes, tail fuses (mounted in the rear of the bomb casing) were also used and their arming was controlled similarly.

The flight suits we have on were the dark olive green NOMEX flight suits.

The yellow baseball hat I have on was my "lucky charm." It flew with me on every combat sortie I flew in SEA and brought me back to a safe landing every time --- as well as A.C. Mizner's skill.

Don Craig had moved over from being a Special Ops AC-119 "Stinger" copilot flying in SVN and Cambodia to BUFFS. He was and IS my closest "buddy" within our crew.

The author before a bombing mission.

Linebacker II – A Personal View

Vince Osborne

I often ponder if I a Vietnam veteran in the true sense of those words? I see the war footage and I know that I didn't have it as tough as those who fought close combat in the jungles. I was also lucky that I never saw any of my friends or fellow fliers blown up. In my crew compartment as an Electronic Warfare Officer on the B-52, I was surrounded by racks of electronic equipment and didn't even have a window to the outside world. I did get a sinking feeling though, when I heard that an airplane in a flight ahead of my B-52 was missing and believed shot downed. It happened on December 18, 1972.

Earlier that day only the pilots were ordered to attend a meeting in the briefing room where we all formed up prior to each bombing mission. When they came out from the closed door meeting they had been ordered not to talk about anything they had been told. It didn't take long, however, for the rumors to spread that we were going to bomb Hanoi. As far as we knew, no B-52s had ever been that far north during this conflict. I personally had flown as far into North Vietnam as Vinh. That had taken place in October, and that was the first time aircraft in my flight had been fired on by SA-2 Guideline missiles. I didn't know for sure and had my doubts that the pilots had been told we were going to Hanoi, so I placed a 50 cents bet of with our pilot that "they" wouldn't send the B-52s there.

Later in the day, as we prepared for our scheduled mission, we donned our flight gear. We climbed into the old K2B flight suits, and gathered the rest of our bags of flight gear, and brief cases of papers and books, and boarded the blue busses that took us to the briefing room—the same room the pilots had occupied in their secret briefing earlier in the day.

As I boarded the bus, I thought back to the navigator, John, on the crew with whom I had been to Southeast Asia (SEA) in 1970. I remember the profound comment he made one day. He said, "I don't have nightmares of getting shot down in an airplane, I have nightmares of blue busses coming to get me."

One of the first items in the briefings we attended before each mission was the showing of a projected transparency of a map with the target outlined in a small triangle. As I looked upon the screen, my eyes went to the area on the map known as the Demilitarized Zone (DMS). From there my eyes scanned gradually north on the map until they came to rest on the triangle showing the target for the upcoming mission. It was Hanoi. Without a word spoken, I unzipped the little pocket on my left sleeve where I kept my change and ballpoint pen and handed the pilot two quarters. I later heard that the senior staff conducting the briefing had planned into the brief some extra time for cheering and applause at this point in the briefing. Unlike the dramatics we grew up watching in the World War II movies, there was dead silence. While I'm sure many of us had long thought that the correct military approach to this war was to try to win it, we also must have understood the true danger of flying a huge airplane into the most strongly defended airspace the world had seen to that point in history.

There was always a place in each pre-take off briefing for a chaplain to say a prayer. The variety of prayers offered was remarkable. Most of the time the chaplains prayed for our safety. Some prayed we would bomb the Godless hoards of communism off the face of the earth. The prayers must have been heard, and for the most part we stayed safe. (It took Reagan to get rid of the Godless hoard, and he did it with economics not bombs.) In addition, on this day and the 10 days that followed that made up what SAC crews came to call "The 11-Day War" or the more formal name of Operation Linebacker II, the Chaplains came out to the flight line to chat with us as we preflighted the aircraft. That was the only time I ever saw this happen. If we had ever thought that this was going to be a routine mission that thought began to dissipate as this singular departure from the normal events occurred.

On the way to the aircraft, my crew stopped by the Personal Equipment (PE) shop and picked up our CHAP kits, as we did before every combat mission. I'm sure I once knew what CHAP stood for, but its meaning is now long gone. These were the net vests that held the survival equipment that augmented the normal survival kit that we sat on. The seat survival kit was left permanently in the aircraft (an integral part of the seat) and was for all practical purposes a peacetime survival kit. It would deploy on a lanyard below us as we departed our ejection seat if it was necessary to bail out of the aircraft. Probably the most important piece in the devices contained in that CHAP kit was the survival radio. One would use it if downed in a combat area and be able to talk to Search and Rescue (SAR) aircraft to coordinate an extraction from the area. I felt the gun in the kit was also important. On peacetime flights we did not carry weapons, but during real combat missions we were each issued a Smith and Wesson .38 caliber 6-shot revolver and 12 rounds (I think) of ammunition. This time I knew the mission was not going to be one of the so called "milk runs" we had flown over South Vietnam before. This was the first time I ever actually loaded the gun before sticking it in the holster of my CHAP kit where it would press on my left kidney for the period of time I wore the kit. We usually waited until we were only about a hundred miles off the coast of Vietnam before we strapped on the uncomfortable vests. Because this vest was worn in addition to the parachute and water wings, it was a lot of equipment to be wearing while trying to operate aircraft equipment.

I made a personal tape recording of quite a bit of my second trip over Hanoi. It's now very interesting to go back and listen to it. On the tape you hear the gunner, Tom, calmly inform the pilot, John, that there are two SA-2 missiles headed toward us. If Tom's voice had gone up a couple of octaves John would probably not have had to ask Tom twice if he was talking about missiles. But he did ask. Crews were briefed prior to the mission to count the number of missile firings, and keep track of the locations from which they were fired. This was just part of the flight information we were required in the intelligence debriefing we attended on the ground after the flight. On the tape the co-pilot says, "I've counted 26 I'm going to quit counting." Also on the tape are the undeniable sounds of the target tracking radar used to highlight our bombers to the ground defense systems, since the audio of our warning receivers was also piped into our intercom system.

Our flight made it through our bombing run without damage, but in doing so, we lost track of the other aircraft for a short period of time. This was right after bomb release and when we were coming off the target. Not knowing where the other bombers were was just as dangerous to the crew as the missiles that were being fired at us. Our pilot elected to turn on our aircraft's rotating beacon for a few seconds so the other two bombers could see us and allow them to get back into formation. I never knew why they didn't see us on their bombing-navigation system (BNS) radar, but things like that happen in the fog of war.

For a long time after that flight, I would get chills when I listened to the tape—literally chills with the hair on my arms standing up. Now, 32 years later, I am finally at a point where it no longer has that effect on me. I seem to look back on that night and see a young man who was a little scared, but did the job he was required to do. I see a young man I don't really know today.

The "Door Rester"

Vince Osborne

I remember one time when I was flying a bombing mission out of U-Tapao Thailand. It would have been in 1972 because the crew I was with that flight was the one with whom I was later to fly with during Linebacker II. To understand the story one needs to know that there is a device in the Radar Bombardier's (RN) compartment called the BRIC (bomb release indicator something). No matter what its official name was Crewdogs just called it "the brick." On its panel is a light that is illuminated to indicate each bomb carried by the B-52D. As the bombs are released, a quick glance by the RN would tell him if the bombs had properly left the aircraft as they were supposed to do. On this particular night one of the lights stayed lit, indicating a bomb had hung in the bomb bay.

The normal checklist procedure for this situation was to have the Navigator enter the bomb bay to see if, in fact, there was a bomb hanging in the rack or if it was just an improper indication on the BRIC. This would occur as we descended below 16,000 feet because the crew compartment had to be depressurized so the door to the deck behind the lower crew compartment could be opened. This door allowed access to the bomb bay. The Nav was required to wear a parachute and carry a "walk-around" emergency oxygen bottle. These bottles were about six inches in diameter; about 20 inches long and had a connecting valve on one end with a pressure indicator dial to show how much oxygen remained in the bottle. The bottle also had an alligator clip that was supposed to make it possible to clip the bottle to the flight uniform and carry it around conveniently. Conveniently? Imagine a man with a 40-pound parachute, carrying the oxygen bottle, carrying a flashlight, bending almost double to get through the little crawlway to maneuver back to the bomb bay. Buzz, the Nav on this flight, did it. He confirmed to the rest of the crew that a 500-pound bomb was actually hanging in the bomb bay.

Here's why it was good for us to know this fact. There was no way to be certain that the fuse didn't spin up and arm the bomb while the bomb bay doors were open for the period when the rest of the bombs were being released. For that reason, the crew had to presume that there was a possibility that the 500 pound bomb that was hanging a few yards behind them (several yards in front of the gunner) was armed. Because of this possibility, a very cautious landing was in order. Since the button had been pushed to release the bomb on the bomb run, it was not certain how stable it was sitting in the rack. A hard landing could jar it enough to make the bomb fall out of the rack and land on the bomb bay doors. This led to the necessity to make a very slow deceleration to a full stop, since a rapid deceleration would mean the bomb would slide forward like a loose cannon and the fuse would contact the front of the bomb bay.

As a reminder, we all had to assume that it was armed. When our wheels touched down the BRIC light went out. We knew the bomb was on the doors. The RN called the crew instantly to let us all know its staus. The pilot told the gunner he would not be deploying the drag chute. The large drogue parachute was used to slow the aircraft down after landing so that a great deal of breaking could be avoided thus making the brakes last longer and not get so hot as the aircraft decelerated to taxi speed.

After touchdown, the pilot slowed the bird gradually using the entire runway (probably about 12,000 feet of it). When we came to a full stop, we didn't taxi anywhere else. We were met by many fire trucks with red lights flashing and the EOD (Explosive Ordinance Demolition) guys whose job it was to enter the bomb bay and put a pin in the fuse of the bomb thus making it safe.

I was the Electronic Warfare Officer (EWO) or EW on the crew. Among my many jobs was to obtain (by signing for), and distribute the classified code materials that might be needed on each combat mission. After each mission I had to account for every page of this material by turning it in and retrieving my signature sheet that had been used when I signed for the material. (Few things could cut a career short faster than a security violation, so losing classified material was a biggie.) As my crew rapidly evacuated the airplane which was very dark, I went busily about with my flashlight trying to find where the various crew members had put their classified paper that I had signed for.

I had some of it in my hand when I heard a voice in the darkness. It was the pilot back at the entry hatch. He said, "Vince, get out of there. We have a bomb on the doors." Well I knew that, and I thought that if it hadn't gone off by now it wasn't going to, and I didn't want a security violation. Despite my doubts, I dutifully got off the airplane, and came face to face with the fire trucks. I scrambled around and finally chose a pair to duck between. I thought to myself, "Getting out of the way of the fire trucks is more dangerous than flying the mission was".

Pretty soon we got the "all clear" and went back inside the airplane and retrieved all our flight gear. I didn't get a security violation and we didn't get blown up. The aircraft was towed to a parking place for the night. I never found out how they got the bomb off the doors. I feel that it must not have been easy, but, being SAC guys, I'll bet there was a procedure and a checklist for it.

Crew of Ruby II being debriefed in Guam, January 1973. (Right-Left) Lt. Col. Gerald F. Wickline, Capt. William F. Milcarek, Maj. Roger A. Klingbeil, Capt. William E. Ferguson, TSgt. Carlos S. Killgore. Missing from the photo is Capt. Myles J. McTernan (Still in hospital.)

The Crew of Ruby II

Pilot
Lt Col Gerald F. Wickline

Navigator
Capt Myles J. McTernan

Co-Pilot
Capt William F. Milcarek

Electronic Warfare Officer
Capt William E. Fergason

Radar Navigator
Roger A. Klingbeil

Gunner
TSGT Carlos S. Killgore

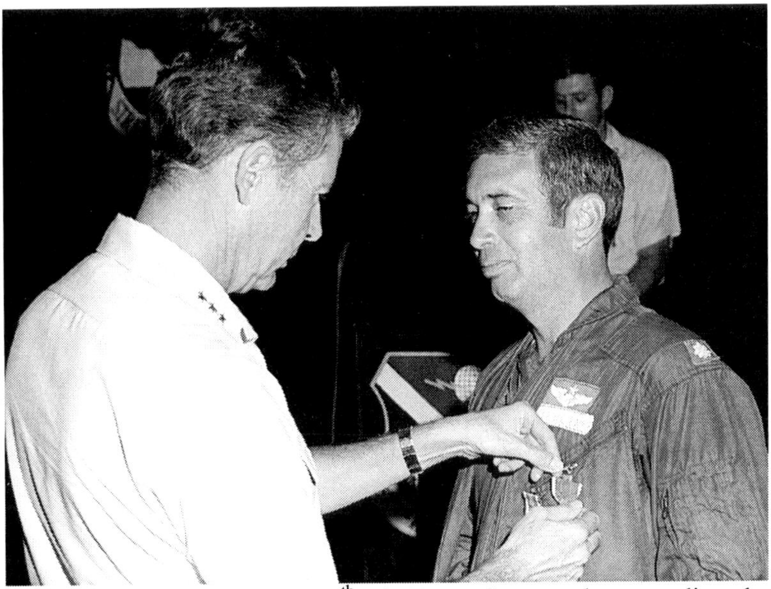

General Gerald W. Johnson, 8th Air Force Commander, awarding the Distinguished Flying Cross and Purple Heart to Lt. Col. Gerald F. Wickline.

Bailout

Jerry Wickline

(The following is a copy of a letter I sent to a friend describing my two missions during Linebacker II, AKA the Eleven Day War, and of our final mission over Vinh on January 3, 1973. Our plane was the last B-52 to be shot down in the Vietnamese War.)

Dear George,

I guess you would be interested in my WAR STORY, so here goes. I deployed to Guam on 9 Feb 1972 and was back home for a 26 day break in July 1972, then back over for five more months. Then another 28 day break at the end of November 1972.

In April 1972, we flew five missions in the DMZ area and southern NRVN where SAM's were fired at us with none coming very close. The rest of missions were flown in South Vietnam, where we were relatively safe. I deployed back over to Guam on 24 Dec 1972, crossing the International Date Line on Christmas Day (I almost missed Christmas completely), and flew our first mission over Hanoi on the 27th. Believe me George, you can't believe how bad it was. Our target was a railroad yard on the north side of Hanoi. We came in from over the water just north of Haiphong, and we no more than coasted in over land when the first missiles were fired and they continued in barrages for the next 23 minutes until we exited the SAM rings near the Laos border. We saw well over 100 SAM's fired and more than 30 of them came within one mile of our airplane, with at least half of those missing us by less than 200 feet. My mouth was so full of cotton I could barely talk, and the whole time, I thought I would be dead the next second. There was no way to outmaneuver them, and several times I was blinded by a near missile detonation or from the brilliant glare of their rocket trail as they went past me. The B-52 right behind me was shot down, with no survivors listed as of this time. The total time of the mission was 16 ½ hours with one pre-strike and one post-strike refueling.

We had 16 hours on the ground and flew our second mission on December 29th. This time we were hitting Phuc Yen airfield (NW of Hanoi) and flew practically the same route. This mission was a milk run compared to the one before. Only about 40 missiles were fired and only about eight came within one mile of us, and none very close. We did have a MIG close to within 3 miles at our 12 o'clock, but he never locked on or fired any missiles. We thought he had the first plane in our cell, because he was only about a mile from him when he broke off the attack. On this raid, we had three waves of B-52's striking the same target, from three different axis - SIMULTANEOUSLY. We were more afraid of another B-52 dropping bombs on us than we were of SAM's and MIG's because we were at 31,000 feet and the lowest altitude of the three waves. Anyhow, no one was shot down or even hit that night, and that was the last raid over Hanoi. This mission was only 14 hours long and as soon as we landed, we were shoved on a KC-135 to go to Utapao-so Christmas in Guam and New Years in Thailand.

We got our training at U-Tapao and flew our first flight on the night of 3 Jan 1973. Our target was a truck park 15 miles NW of Vinh, which is just below the 20th parallel. We were the next to last airplane in a nine plane wave. Our call sign was Ruby 2. At about the IP we picked up some SAM signals and started our SAM maneuvers. We saw lots of AAA fire over Vinh, more that I have ever seen before, including the two missions over Hanoi. Several missiles were fired at planes ahead of us, and at the other two waves of B-52's which were hitting targets West and SW of Vinh.

At about 40 seconds TTG, four missiles were fired at us, off our left wing. SAM's #1 and #2 were fired first followed a few seconds later by #3and #4. I maneuvered to avoid #1 and #2 with #1 going past our nose (real close) and detonating just above us (going up). #2 SAM missed our tail by an estimated 50 feet and also detonated just above us. I had lost sight of #3 and #4, and we were close to TTG zero so I rolled level with the PDI centered and simultaneously with bombs away, there was a tremendous explosion directly below our nose. Three windows on my side of the airplane shattered and showered us with broken bits of glass. The #1 engine fire light came on and a blazing fire was burning in the #1 pod. We shut #1 down. All of my flight instruments, including airspeed and attitude were inoperative. Most of the engine instruments on both sides of the cockpit were inoperative, in fact the glass was shattered in most of them. All hydraulic power to the left wing was out and all fuel gauges on the left wing were either spinning or stuck.

I polled the crew and everyone answered but the tail gunner. No one was hurt at this time, but a couple of minutes later, the Radar Navigator called in a fuel leak that had just developed above his head, and fuel was pouring in, right on top of him, and shortly he started screaming from the pain of the fuel burning him (chemical burns).
We stayed in cell with the other two airplanes and three more SAM's passed across our nose and exploded just above us. We got out over the water and turned south and by this time the RN was screaming from the pain of the JP-4 burns, and the lower deck was literally floating in JP-4. The escape hatches were nearly brim full of fuel.

We tried continuously to contact the gunner, but to no avail. We started a descent with the other two airplanes staying with us, and leveled off at 12,000 feet about 90 miles north of DaNang. Soon after level off, we felt a thump and heard a parachute beeper go off. Our gunner had bailed out all on his own, as flames from a fire in the belly began licking around his turret. That was one of the smartest things the gunner ever did, as had I not heard from him I may have tried to land the airplane and would surely have killed us all. Once I was sure the rescue forces were in contact with the gunner, we started setting up for the bailout.

The airplane was barely controllable. Every time we would attempt to slow down, the airplane would start a roll to the right and we could only straighten it out by gaining airspeed. The fire in the #1 pod kept burning intermittently. The #8 throttle had no control over the engine and I think it was running at idle. As near as I could tell, the other six engines were still running, but I was not sure because I had very few reliable engine instruments.

By this time we were abeam DaNang, so I turned eastbound (away from land) and ordered bailout. We were almost sure that when the first ejection seat fired we would become a fireball, but thank God that didn't happen. I heard two thumps which I assumed was the Navigator and RN ejection seats firing, followed a few seconds later by a third, which was the EW. I told the co-pilot to go and he squeezed the trigger. I was temporarily blinded by flying debris. After I got my eyes cleared and looked, he was gone. I made a call on interphone to see if anyone was still on board – waited a few seconds – pulled the throttles to idle – assumed the position, and squeezed the trigger. I felt a tremendous kick in the seat of my pants, a great blast of cold air, severe tumbling, an extremely sharp jolt tearing the ejection seat from my hands, and a loud pop followed by the most intense silence I have ever heard. This all happened in seconds as my parachute zero delay lanyard was hooked.

I could hardly believe my good fortune when I looked up and saw that beautiful big orange and white canopy of the parachute above my head. I said out loud to myself, "Wickline you lucky son of a gun, you've got it made now." About that time I felt extremely nauseous, so I pulled my oxygen mask from my face and barfed into to the Gulf of Tonkin. I think it came from the relief of the last 30 minutes, and that I was away from that airplane, which was a fireball waiting to happen.

It took about 15 minutes or so to come down from 12,000 feet in the parachute and it was so quiet that it almost made your ears hurt. I saw a fireball on the horizon. Our B-52 had hit the water and exploded. I started to oscillate in the parachute, swinging from side to side, so I reached up to grab the risers to stop the oscillations and could barely get my right arm above my head, and when I did, it hurt like hell. I had to do this three or four times during the descent, and each time my right shoulder hurt worse and by the time I hit the water, my right arm was useless and very painful. I started hearing the sound of waves, and as I had already deployed my survival kit, I uncovered my parachute quick disconnect covers and got ready to hit the water. As my feet hit the water, I released my parachute and barely bobbed below the surface. I rested a few minutes in my life vest and then tried to get into my life raft, which was a major accomplishment because the waves were 10' to 15' high and white capping, and about every third wave would wash completely over me. My right arm was hurting real bad by this time and it was practically useless, so it took me about 15 minutes to get into the raft. When I did get into the raft, I found that I had planted my rear end in the small end of the raft, instead of the large end, which made the raft very unstable. I was not about to get out and get back in the raft the right way, so I just hung on. I tried four or five times to get out of my parachute harness, but couldn't, so I just sat back, got my survival radio out, turned it on and waited. I heard nothing for a while, then I heard my co-pilot talking to an OV-10 pilot so I knew they were looking for us. About this time the sun came up, and with it the wind, which made the waves higher and chilled me to the bone. There was a low overcast and lots of haze, with very poor visibility. The co-pilot continued to talk to the OV-10, giving him short counts for ADF and soon after, I heard an engine, and looked up to see an OV-10 approaching. He passed over me first, so I gave him a call and he set up an orbit directly over me, and started to direct the helicopter, which was already airborne, to the area. About an hour after I hit the water, I saw the helicopter, and watched while he picked up the co-pilot who was only about 200 yards from me. Then he came over and picked me up.

Even though my arm hurt like hell, I attached myself to the rescue devise with a deathlike grip and was hoisted into the helicopter. We were flown directly to the DaNang hospital.

My co-pilot was in good shape except for a sore back and a cut under his chin. The gunner had been picked up by a helicopter from the USS Saratoga and was OK except for a cut under his chin. As a matter of fact, all six of us had a ½" to a 1" cut under our chins. The RN was next to be picked up and was blistered badly about his waist, wrists and ankles. (Anyplace his clothes fit tightly) The EW was next, with a sore back and a piece of helmet visor, which shattered during the ejection, in his right eye. There was no permanent damage.

We had a lapse of a couple of hours with no word from the navigator. Then we got the word he had been picked up in real bad shape, with a large cut on his right cheek, a fractured cheekbone, and a distended stomach from swallowing so much seawater. His ejection seat had failed to fire and he made a manual bailout, and probably hit the bottom of the airplane on the way out. He is near full recovery now and should be back to flying soon.

They took x-rays of my shoulder at DaNang and found a displaced fracture of the large bone in my upper arm. A piece of bone, at the joint, about the size of a half dollar and about ½" thick had broken loose and was pulled in behind my right shoulder. There was no immediate need to have surgery, so I was transferred to Guam where General Johnson awarded me the Distinguished Flying Cross and a Purple Heart and was then air evacuated back to Dyess AFB. The next day I was sent over to Carswell AFB, and the following day they made a 5" incision in my right shoulder, replaced the large piece of bone that had broken loose, and secured it in place with a 1" screw, which will stay there forever unless it gives me problems. I have gone through three months of therapy and still can't get my right arm above my head. It looks like that is all I'm going to get. It still causes me some pain, especially with a lot use or heavy activity that causes me to move my arm to limits.

Well George, that is the story. I am thankful to be alive. The mission over Hanoi on December 27, 1972 was a nightmare I will not soon forget. I found out on that mission that I was not a coward. I begged for an excuse to turn that airplane around and not fly through those missiles, but I was more afraid of being branded a coward, than of dying.

Your friend,
Jerry

IP Inbound – The Final Minutes of Charcoal 1

Robert G. Certain

(Author's Note: Charcoal 1 was a B-52G 58-0201 and was manned by S-18, a select crew out of Blytheville AFB, Arkansas on 18 December 1972, the beginning day of Linebacker II. This account of the bomb run is from *Unchained Eagle: From Prisoner of War to Prisoner of Christ* by the crew navigator, Robert G. Certain.)

As we made our final turn toward the target we entered the 90-knot jet stream, boosting our groundspeed to 560 knots. At least our time in the area was being reduced by a few minutes, and our groundspeed might make it more difficult for the SAM crews to hit us at least until we made our steep, long post-target turn to the west to exit the area. That turn, which would blank out our SAM jammers for nearly two minutes and then reduce our groundspeed to about 400 knots, was another insane feature of the worsening tactical situation.

The EW, Capt. Tom Simpson, reported 100-millimeter anti-aircraft artillery (AAA) radar was locked on, but would be ineffective at our altitude. What we didn't know was that our ALT-22 Modulated ECM Transmitters could be countered by the SAM crews. The D models and some Gs had ALT-6B Unmodulated Transmitters that were more effective.

With our outside radios still off and the crew maintaining only checklist and bombing instructions on the intercom, the Radar and I were able to concentrate on this critical offensive phase of the mission. By this time I was aware of no emotion other than dogged determination, no words other than checklist items, and few thoughts other than prayers. Thirty seconds before bombs away, Dick opened the doors, and twenty seconds later I was to restart my stopwatch as a backup to the drop should anything go wrong. Just short of ten seconds to go, time seemed to stand still and speed ahead, all at the same time.

At 1313 GMT, the radar screens went blank and other instruments lost power and the aircraft shuddered and yawed slightly left. My first thought was that the copilot had accidentally knocked the generators off line. Before I could speak, though, he was shouting over the intercom, "They've got the Pilot! They've got the Pilot!" I thought, *Who has the pilot?*

The EW was also shouting, "Is anybody there? Gunner, gunner!" His cockpit had gone black, his equipment had major electrical shorts and explosions, and the gunner was covered in blood and slumped in his seat. Tom had also lost his earphones but not his microphone. We could hear him, but he couldn't hear us. Dick and I had intermittent intercom.

My internal voice was saying, *This can't be me. We haven't been hit; or have we?* Then the voice silently whispered, *Yes, this is for real.* Then, that last internal "still, small voice" took over, and everything went into slow motion. From that point forward, I was truly on auto-pilot, reacting in the way I had been trained. That training was for the worst-case scenario, and now all of a sudden, we were cashing in the "training-chit" with every last ounce of our energy! We are performing and reacting on autopilot. Thank God it worked!

I pulled out my small Sanyo flashlight to check the Altimeter and noted we were in a 400-500 feet per minute descent. I then laid it on my desk to check my personal equipment; and Dick picked it up and aimed it upstairs toward Tom. Tom's voice could be heard briefly, "Is anyone alive up front?" It was then very quiet, and I suspected we had lost the 4 port engines, and maybe more.

The first SAM had exploded to the left front of the aircraft, scattering hot shrapnel like a shotgun blast at a velocity of 8,000 feet per second. The plane flew at 490 knots true airspeed through the burst, and thousands of pieces of hot metal shards ripped through the plane, wounding the pilot and killing the gunner. Some of it was sucked into jet engines (which destroyed them and the engine-driven generators), and others cut hydraulic lines and set the oil on fire. We probably had ruptured fuel tanks on fire, but there would be no major explosion unless the JP-4 lit off in an intact tank. That might come in a few seconds.

Over the Gulf of Tonkin, my brother, Capt. John Neal Certain, the pilot of a Stratotanker in the refueling orbit, saw the SAM hit the B-52 and the explosions and flames as it began its death spiral toward the ground. He knew I was to have returned home that morning, but a sick sensation in his abdomen told him that I was probably in the middle of that conflagration.

I looked over my left shoulder and saw a fire in the forward wheel well through the porthole in the bulkhead door behind the offense cockpit. First I thought of the twenty-seven 750-pound bombs in the bomb bay right behind the fire, and turned to the Radar, "Drop those bombs!" He safetied them (we didn't know where they would land, so we did not want them armed since we were under orders to avoid POW camps and civilian areas), and hit the release switch. They all seemed to drop away from our now-crippled BUFF. Oddly, though the indicator lights went out, I felt no shudder or lurch as they left. My next thought was that the fire was also directly below the main mid-body fuel tank, loaded with 10,000 pounds of JP-4.

"The pilot's still alive," Don's voice came weakly over the intercom.

"Co, Nav, escape heading is 290." The Tonkin Gulf 60 miles to the east would have been a better place to go for bailout; but there were too many SAM sites and MiG airfields in that direction.

"EW's leaving!" In a controlled ejection, the navigator went first (to make a hole for manual bailouts if any other seats failed), then the Defense Team, the copilot, the Radar, and finally the pilot. In an uncontrolled escape, it was every man for himself. This was definitely an uncontrolled event!

With that, about ten seconds after the first of two SAMs hit the plane, the first crewmember ejected. I heard the explosion of his hatch and seat as it rocketed up and out, but felt no decompression. We had been at combat pressurization (17,000 feet), but the holes in the airplane had caused that to rise to over 30,000 feet by the time we ejected. I looked at the Radar. Our eyes met, and we both started preparing for ejection. I threw my flight case as far to the rear of the cockpit as I could, cleared my table, tossing pens, dividers, and charts to the back (I didn't want any of that stuff being drawn out the hatch when I ejected, possibly causing injuries), stowed my table, pulled down my visor, cinched up the oxygen mask, pulled the "green apple" to activate the flow of O_2, kicked back into the leg restraints. I discovered the restraints had already tripped; so I had to bend down and reset them before kicking back again. I grabbed the ejection handle between my knees – incorrectly, with palms facing me. Instantly, I remembered that my elbows would extend past the hatch opening if my hands were placed that way, reversed my grip, looked at the Radar again, turned to face forward, saw the ejection light come on as the pilot ejected, and pulled. I thought the seat had failed. The ballistic activators were supposed to blow the hatch below my seat and fire me in the seat out the bottom of the plane, all in three-tenths of a second.

But fear gripped me and expletives filled my mind as the panels in front of me seemed to move up exceptionally slowly. *Was this thing jammed on the rails? I'll be beaten to death by the wind!* It was an illusion brought on by the state of shock induced by the extreme danger of being in a burning and likely to explode bomber directly over the enemy capital. Everything from the first call from the copilot to ejection had appeared in exaggerated slow motion, like a movie depiction of an automobile crash. In actuality, it was probably about 10 seconds.

The next thing I knew, I was in the cold air of the troposphere. At 31,000 feet, the temperature was –55° centigrade. As I tumbled around all three axes, I thought, *That was a dumb thing to do. I'll bet the plane was still flyable. Where is it? Perhaps I could crawl back in.* Dumb thoughts from a lingering sense of invulnerability, I suppose. *Now what do I do? OK, God, it's you and me.* My prayers for the next few minutes would remain blunt and somewhat profane.

Next, I checked to make sure the 350-pound seat had separated from me. It had. One second after clearing the hatch, more ballistic activators had blown seat belts and shoulder harnesses loose and activated the man-seat separator, a strap on an inertial reel running behind and below me. When it tightened, I was pushed out and away from the seat, arming the parachute for automatic deployment passing 15,000 feet. I carefully checked to feel my parachute harness. Had I remembered to connect all the clips? I had only checked it about two dozen times on the way in. It was there. Then I concentrated on stabilizing my fall. My oxygen mask was being forced up over my eyes. I pulled it down to my mouth, tightened the connections, repositioned my visor and tightened the lock nut. But still, the force of the tumble kept pushing it up. My mind was racing through checklists and training manuals and all those boring safety lectures I had sat through for the last four years. I was amazed and calmed that I seemed to be remembering so much.

I tried to get into the sky diving position to stabilize in a facedown fall, but that only made matters worse. I doubt that I tried that for more than a few seconds, though it seemed longer. Grabbing my mask, I held it in place with both hands, pulled my legs together, bent my knees, and bent over slightly. Right away, the tumbling stopped as my personal center of gravity was shifted to my lower body, and I was falling feet first toward the earth, now about 20,000 feet below. I began to look around for signs of Charcoal One. Above me I could see a long trail of smoke – wide, thick, and dense – turning slightly west, then looping around and descending before disappearing into the clouds below.

I could feel the effects of the wind, but I had no sensation of falling. I slipped two fingers of my right hand into the T-handle to manually activate the chute should I enter any clouds (briefed to be solid at 2,000 feet). *Just wait. A high-altitude opening shock could damage the canopy and injure the jumper.* After ejection I was traveling initially at the airspeed of the plane. As I freefell I slowed to terminal velocity (when the wind resistance is equal to the force of gravity, about 120 knots). In a moment, I felt the parachute opening as I passed through 15,000 feet and fully deploy a few hundred feet below. I felt each fold, one at my shoulders, one at my buttocks, as it began its deployment out of the pack. So far, all my emergency equipment was working properly.

With the opening shock I was swinging like a pendulum in a wide arc below the canopy. First, I checked for a good and complete chute, then reached up and pulled the red daisy-chained line on my right riser to release four lines and create a small spillway. That accomplished, the chute stabilized, and the undulations ceased. Finally, I looked down for the first time. There, between my boots, I saw the inferno that made up the three targets that had been struck over the last twenty minutes. As I watched, I saw a series of explosions walk though the target – the twenty-seven bombs from Ebony Three had found paydirt. The time was 1315:50 when Ebony Three's bombs started exploding. I had been falling for two minutes and forty seconds.

Was that the last of the bombs, or were there others falling all around me? Am I really going to land in the target and be burned alive? Grabbing two fists full of right risers, I pulled them down to my ankle, tilting the military canopy and steering it to the west. My heart was racing as fast as my mind. Somewhere around 10,000 feet, the wind shifted out of the east, aiding in moving me west and away from the target area. This nightmare was going exceptionally well.

The solid undercast turned out to be broken to scattered clouds. The moon was full, and I could see the ground clearly all around. Panic was beginning to replace concern. White panels in the canopy, and a white helmet were not going to be assets as I slowly descended into enemy territory no more than ten kilometers northwest of Hanoi. I was going to have to find a wooded area to land in if I was to have any chance of evasion. I looked around the sky above and below for other chutes, but spotted none. Somewhere out there were Tom and Don for sure, and I hoped Dick, Bobby and Fergie were somewhere above me.

As I was beginning to feel relief drifting about thirty degrees west-northwest away from the target, in my right peripheral vision I caught another series of explosions – right in line with my drift. *Oh, God, now what? There shouldn't be another target over there; that was our escape route.* As I looked down through a break in the clouds and drifted past this new blaze (and through the mushroom cloud of smoke), I realized that this fire was shaped like an arrow – Charcoal 1 had plowed in flames into a field below at approximately 21.07N, 105.52E. My watch showed 1317 GMT. Ejection hadn't been such a bad idea after all. *Look Lord, it's you and me. If I'm gonna die down there, just go ahead and let me die right now. I'm ready to go. I would just as soon not have those people down there get me and kill me. I don't care if they capture me. But if they're gonna kill me, you take me now.*

The situation was becoming clearly hopeless. We had been told back on Guam that search and rescue would not be available. The apparently populated farm country below me offered no place to hide. I had heard horror stories of how wounded aircrew had been killed or left to die, with medical care being offered only if they lived for three days. Being captured with a major injury was not a pleasant thought. I needed to make my fall a good one.

I had some static ground training in the harness, a few dozen parachute landing falls from a platform, and two parasailing rides behind a pickup truck. The cable was about 300 feet long, so we were towed up about 200 feet, cut loose and drifted to the ground. The theory is that falling the first 30,000 feet doesn't hurt – hitting the ground does the damage. The big thing to learn is to land properly. I hoped I had learned it well.

I opened the 60-pound seat kit to string out all of the survival equipment it contained on a tether below, and to decrease my chance of injury on landing. Then I faced forward keeping my eyes on the horizon, grabbed my risers, put my legs together with my knees slightly bent, and waited. When my toes began to touch the earth, I executed a parachute-landing fall (PLF), rolling to the right into a dry ditch. The point of touchdown was approximately 21.07N, 105.50.5E. To the east side was a plowed field, to the west side a railroad. My injuries to this point consisted of a few bumps, bruises and abrasions from the ejection, fall, and landing. The PLF had gone perfectly. Everything continued to work according to the book. But there was no time to relish in this small success. It was time to evade.

(L to R) CP left rear, Jim Hooppaw - Pilot and author, RN, N, G and EW of Crew E-13, Clinton-Sherman AFB, OK. Picture was taken in U-Tapao, Thailand in the spring of 1968.

The Dreaded Bag Drag

James Hooppaw

(The following is an excerpt from "*Where the BUFF Fellows Roamed*" by James Hooppaw. It is reproduced with permission of the author. He has also written "*Burns Flat Motoring and Yachting Societe`*" and *"Tall Tail Tales".)*

The Arc Light mission was not shrouded in secrecy, but those who were not participants may wonder how it was completed. Very simply, each mission was made up of two cells of three aircraft called a wave. There were four to six hours between each wave launch. The two cells were separated by twelve to twenty minutes. The aircraft in each cell would takeoff at one-minute intervals. This was changed later when all flights were flown out of U-Tapao, Thailand. A wave consisted on only one cell made up of three aircraft.

Each crew would preflight their assigned aircraft. A spare crew would preflight an aircraft that was designated the "manned spare."

This crew would start engines and be prepared to fill any position that might become vacant due to mechanical problems. It also taxied to the runway and was available within certain time frames to fill in if any airborne aircraft aborted. There was also an unmanned spare that would be pre-flighted and cocked in case the spare was used and another bird fell out of the launch. Sometimes this occurred before the first aircraft started engines, which was not too bad. Occasionally, it occurred at a time when the response became a goat rope. This resulted in a *bag drag*.

No one liked a bag drag. Things could get lost easily, temporarily, or forever.

Bag drags had been used for many years, but not with the urgency of an Arc Light mission. They became an art form. In reality, they were simple. The crew scheduled to fly one aircraft with a specific takeoff time would preflight the assigned aircraft, and if everything checked out, start engines and fly the mission as scheduled. Everything did not always check out and the mission would be delayed or cancelled. The only thing the crew missed in most cases was a training flight. Normally, there was not a backup aircraft the crew could go to at the last minute and continue.

Arc Light was an exception. We had a manned spare that could replace any aircraft that fell out of the lineup. This aircraft would have engines running and be positioned to be able to fill in at any point in the stream without delaying it unduly. We also had an unmanned spare, pre-flighted and cocked for engine start if needed.

When in the course of human events, something went wrong or an aircraft aborted, the crew, if directed, would bag drag to the spare. This consisted of gathering all equipment the crew had loaded on the original aircraft. This would be put on a maintenance truck (hopefully, just one) with the crew and driven to the spare. This could get hectic when a launch was in progress, and it never seemed to occur at any other time. If engines had been started, it involved shutting them down, gathering the professional gear, secrets box, cooler, chapkit box, lunches and anything else the crew had on board, and racing to the spare. Upon arrival at the spare the pilots would start the engines while the rest of the crew onloaded all of the equipment. It was difficult not to miss at least one thing. The crew was racing the clock to meet a takeoff that would keep them in the cell to which they were assigned, or be able to launch and fill in at the end of the stream within the time constraints of the mission. Time on target was the critical factor.

The crew knew of the possibility, but until we had run the drill we did not really plan ahead. After our first bag drag we did things a little differently. We consolidated as much as we could to reduce the number of items. We did not get comfortable and spread out until we were airborne and on our way.

There were few bag drags when a crew was deploying from U-Tapao or Kadena. It addition to the normal complement of gear, we had personal gear and anything we had purchased. Depending upon the departure base, this could include extensive stereo equipment, china, bronzeware, etc. Naturally, this would not fit into the crew compartment, and the bomb bay was filled, so it was loaded into the 47 section, an open compartment in the fuselage aft of the bomb bay. It contained little but electrical equipment. Unless it occurred early a complete bag drag was almost impossible. Items left might never be seen again. We took birds that, under different conditions would have remained on the ramp.

We were introduced to the bag drag on our second mission. We were the spare crew and would start engines and taxi across the runway and wait, ready to fill any hole that developed. Everything went as scheduled. We went through our preflight. Then, Charlie (an experienced instructor pilot who controlled the launch) called to tell us that number two in the first cell had a broken airplane. They would take the spare we had just pre-flighted and we would move to the unmanned spare. This move was reasonably smooth. The maintenance troops transferred us to the spare. It was time to start engines when we arrived at the unmanned spare. The copilot and I were in the process, while the crew loaded the gear. Charlie called and advised us number three in the second cell had a broken bird and would arrive to take the one we were in.

The crew gathered the gear and loaded it on maintenance vehicles while I sat on the brakes with the engines running and waited for the other crew to appear. They arrived and the pilot team replaced me. We were transported to the original number two which had been declared back in service and ready to fly. We ran a quick check, started engines, and taxied to arrive at the end of the runway just as the last aircraft in the second cell started his takeoff roll.

We paused to recheck those items we had rushed through and began to relax as everything started to fall into place. The copilot and I were the only ones strapped in with helmets and parachutes connected. The rest of the crew were on headsets and just strapped in with the seat belt, for we knew we would not possibly be flying today. Thirty seconds prior to the expiration of our launch envelope the nav advised me: "Pilot, time is almost up. We can relax and get clearance to return to parking." These words were no sooner out of his mouth when Charlie called.

"Spare. We've had an airborne abort. You are cleared for takeoff."

I replied: "Roger, spare copies and rolling. Advise tower that I will be turning left immediately after liftoff. Will call airborne." I was adding power to move into position as I told the crew to strap in and check in. They were ready and checked in by position before I was lined up. The takeoff was normal-the aircraft unstuck and I attained about 200 feet, raised the gear and broke left across the parallel runway. This maneuver saved us many minutes on our departure; we could now raise our flaps on departure heading and not wait to turn until they were up.

I checked in with Charlie, and he advised us we would be number three in the second cell, as the other aircraft had moved up to fill the position of the aborting aircraft. If we could catch up prior to the air refueling, we would get the scheduled offload. The nav team got busy and we set out for the intercept.

Several hours later as the second aircraft in the cell rolled out of his turn into the air refueling initial point, I pulled in behind him and reported in the number three position. The mission was normal from there on until we started home. Since we had burned a lot of fuel to catch up, we had a tanker waiting for us on the return trip to Andersen. We broke off from the cell, hit the tanker and returned to Andersen alone. From that time forward the nav was forbidden to comment on the possibility of launching when we pulled spare duty. All in all, it was a good experience, for we reevaluated the amount of baggage we carried on future flights.

I learned about flying from that!

Dyess AFB Crew E-06 returns from a bombing mission.

Just Another Flight

George Robert Dempsey

It was a dark and stormy night. Okay, it was dark, but that didn't matter much in the Nav position of a Mother Buff, since it is always dark there.

I was back in the war. Our crew had been rotated back home to the states just before the Linebacker II missions, when everybody in a B-52 went to Hanoi and Haiphong. We missed those. We got back to Thailand and Guam just in time for a couple of runs at Haiphong and then President Nixon halted all the bombing in North Vietnam.

That dark night was special. Every Buff at U-Tapao was taking off one after another. My crew was one of those picked for this mission. Our plane was the 3rd aircraft in the 1st cell scheduled over the target that night. After takeoff we went up to the holding box and waited for the other planes to get airborne, since we were among the first group to launch. Our target that night was Vihn. It was the farthest north target still available to the planners and every plane in the night's bombing wave was going there.

Things were uneventful prior to the target, but that soon changed. Just after bomb release, the Radar Navigator on the crew, Elgie Henderson, looked down through the optical bomb sight and called out that two SAM missiles were headed toward our aircraft. About the same time, the co-pilot, Al Hutson, called two more going out in front of the aircraft and the Gunner, Karl Nedela, called two different ones going out in back.

As the seconds slowly ticked by, the pilots watched the missiles turn toward us and the gunner confirmed the same thing was happening with the two SAMs he was tracking. Elgie said the circles indicating the position of the missiles were getting closer.

Our Pilot, Larry Cavendar, was listening to the crew calls, but was holding the plane level and steady. He said, "They have to get closer, then I'll maneuver." Finally, he firewalled the eight throttles and when everyone's voices about the positions of the missiles were getting faster and higher pitched, he rolled our Mother Buff over on the left side on her back and we went straight down!

They say that in terrible times of stress, your life flashes before your eyes. This was not my case. As the plane was diving straight down and I was hanging onto the D-ring of my ejection seat, all that was going through my mind was the voice of Peggy Lee singing, "Is this all there is - to dying?"

Within seconds after the maneuver, the missiles blew up above us and we listened as the gunner Karl called the explosions to the rest of the crew. We lost 9,000 feet in altitude before Larry slowly pulled back on the yoke and brought her back level. The G-forces of the roll-out pushed my butt hard into the ejection seat. That is not a feeling you're supposed to experience in a Buff.

But the excitement was not over for us that night. We slowly began climbing back up to our flight plan altitude. Just after leveling off, Elgie called out over the intercom that six more missiles were coming toward us. The pilot's response to the new threat was about the same as before – same calls, same level and steady flight, and still full throttles. Only this time, when he started the defensive maneuver, Larry turned the yoke and flipped the buff over on the right side. Again, shortly after that, the missiles blew up above us.

We only descended about 3,000 feet on that quick trip. Later, after we were on the ground and were talking over the mission, I asked Larry why he went over the left the first time and over the right the second time. His reply was that he wanted to stress her equally.

When we climbed back up to altitude, we were "feet wet" meaning that we were out over the water and headed east. Just as we thought we were about out of range of the missiles, Karl's voice came back on the intercom calling two more coming at us from the back. Larry asked me how far off shore we were. I replied "20 miles". The SAMs kept coming. At 22 miles, Karl, in his very quiet and calm voice said, "Pilot, I think you should maneuver." Larry knew that meant NOW. He pulled hard and just then, Karl said, "There they go, about one-fourth mile behind us." The missiles exploded without causing any damage.

We started climbing back up to 40,000 feet for the trip south. We knew we were no longer in the cell position we were in before the excitement began. We were back in the middle of the pack after all of our side trips so we went on out a little farther before turning our plane south. As we turned right and were still climbing, I heard a voice yell in my helmet's headset and felt the throttles come back and the aircraft's nose drop suddenly. In the dark of the night, with no exterior lights on any of the planes, Al looked out and suddenly saw the tail numbers on the vertical stabilizer of another Buff beside his co-pilot window. We dropped our Buff down a little more and moved farther out of the formation before climbing again.

As we collected our senses and started thinking about getting back to U-Tapao, Al informed the crew that we didn't have enough fuel to be this far back in the formation for a safe landing. Larry was able to get approval to cut across our route before it was our turn and that allowed us to make up distance and position, so we were able to safely land back at our beloved U-Tapao with sufficient fuel in the tanks.

Later on a flight home in the back of a KC-135, I replayed my experience of the maneuvers on the Vihn mission to a Boeing Engineer that was also going home on the same flight. He sat and listened to my story and when it was over he remarked in amazement, "Wow, she is tough! The wings should have come off!"

Just one night out of 335 nights.

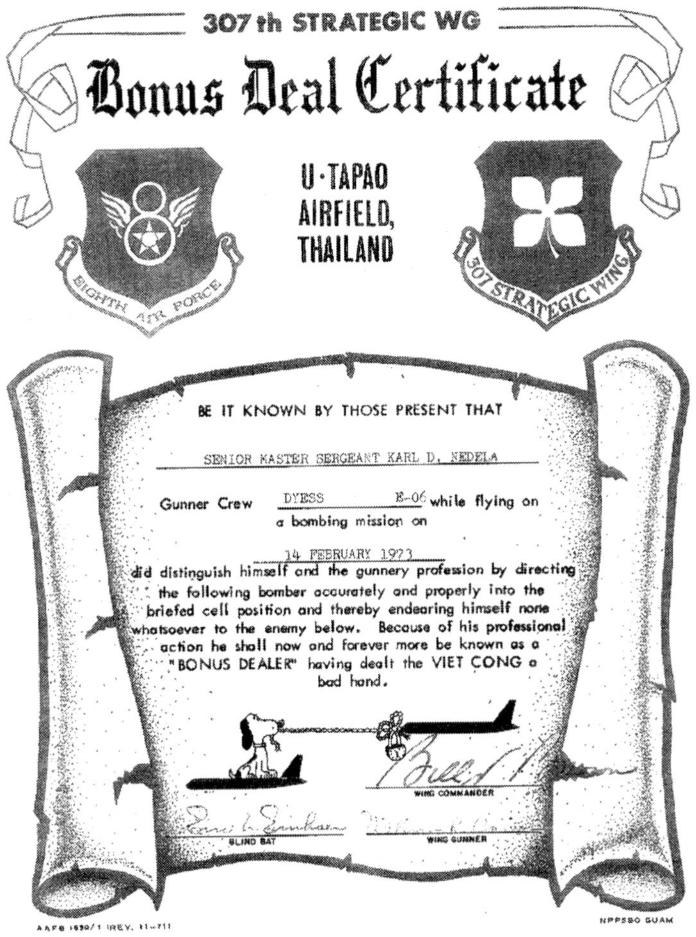

307th STRATEGIC WG

Bonus Deal Certificate

U-TAPAO AIRFIELD, THAILAND

EIGHTH AIR FORCE

307 STRATEGIC WING

BE IT KNOWN BY THOSE PRESENT THAT

SENIOR MASTER SERGEANT KARL D. NEDELA

Gunner Crew DYESS E-06 while flying on a bombing mission on

14 FEBRUARY 1973

did distinguish himself and the gunnery profession by directing the following bomber accurately and properly into the briefed cell position and thereby endearing himself none whatsoever to the enemy below. Because of his professional action he shall now and forever more be known as a "BONUS DEALER" having dealt the VIET CONG a bad hand.

WING COMMANDER

BLIND BAT

WING GUNNER

During the bombing in Southeast Asia sometimes the gunner was required to help during a bomb run. In a BONUS DEAL, the gunner in the leading B-52 aircraft would lock onto the following aircraft with his fire control radar. The gunner would then call range and angle data to help the following aircraft stay on station and deliver the bombs on target.

Houches at U-T

Crew trailer and Officers' Club at U-T

Local transportation in Thailand

Unclassified Material

Outdoor theater in Guam.

Tommy Towery at tent City at Andersen.

Andersen AFB Guam BOQ

The author at a favorite Thai jewelry shop

A Few Shots from the Gunners

Karl N. (Ned) Nedela

Here is a collection of odd and end stories about my B-52 days that I want to share with everyone.

And the Horse You Rode in On…

On December 31, 1972, a couple of gunners at U-Tapao decided to celebrate New Years Eve in rare fashion. This was during Linebacker II missions and things were very tight.

These two men went to the Red Horse Engineers organization and stole their mascot, a small horse. They put this horse on a lead and took it to the Officers' Club. One of the gunners walked into the front door and led the horse across the dance floor.

The commanding general approached and said: "I have never said this before, but, get that @#$%X horse out of this club." The AP's were waiting for the gunner as he left the club and he was arrested. The only questions he would answer were his name, rank, serial number, and date of birth. His pilot got him out and no charges were filed.

In January, the gunners in the MOD got together and bought a beautiful horse blanket and presented it to the Engineers' Organization.

Hooch girls at U-T.

A Meal with Peanuts

I had some fantastic happenings while in Thailand. Flying the Buff from there created some real experiences. We had a maid, Peanuts, in our MOD-5 that was so kind to everybody there. She invited a friend and me to her house for dinner with her and her husband. We rode a Baht Bus near Kilo Sipp where they were waiting for us and led us down a path in the forest to their house. Our shoes were left on the front porch and we entered their home. Knowing our "weak stomachs" they had bottled water for us and Peanuts commenced to fix our dinner. She had two 3 pound coffee cans with wire screen over them and a hole cut in the bottom side to put in the charcoal. Their children had already been fed. Our meal consisted of: noodle soup with pieces of chicken, pork, and vegetables. The main course consisted of fried rice, pork, with cucumbers on top. I don't know if my friend, Luke, or I had ever eaten such a wonderful meal that was fixed with as much pride and care under what we would consider very primitive conditions. A week later, Luke and I invited Peanuts and her family to the NCO Club for a dinner. After retiring from the Air Force I taught High School for 23 years. I told of this experience to every one of my classes and they were amazed.

No Tipping

For those who were TDY to U-Tapao, the O Club and the NCO Club were very popular and very busy. The bureaucrats of Thailand found out that the people working in the clubs were making more money in tips than the "public servants" were receiving for their professional jobs. On every wall in the clubs they posted these signs: "NO TIPPING ALLOWED AT ANY TIME". We continued to do as we pleased but each day there would be one less sign left posted. In a matter of a few weeks, there were zero "No Tipping" signs around. We Americans manage to mess up the economy of many countries but it was done to say "Thanks for the service."

The Must Fly Mission

Once when I was assigned to the 95th Bomb Wing, Biggs AFB, El Paso, Texas, my crew was assigned to a "must fly mission".

When we got to the plane we determined it was not in flying condition because one of the air brakes would not function properly. The wing commander came to the bus where we were waiting and reported that the malfunction could not be fixed for 24 hours so he wanted us to take the plane anyway and fly with the problem because this plane "had to be flown".

Our Pilot, Ben, asked a couple of questions about the dangers of taking the plane with the known problems and could not get them answered to his satisfaction.

This was his answer to the wing commander: "As the pilot of this crew and person who would be responsible for the safety of this flight, we will wait until it is fixed."

Ben was again reminded about the necessity of this on time flight. His final answer to this request was: "Ned, get your stuff out of the plane. Colonel, if you want this plane flown, the keys are in it. My crew is going home."

Did You Bring The Marshmallows?

In the summer of 1956, we were going through B-52 Transition. We were in plane number 8710. After several hours of flying we came back to Castle AFB for touch and go practice landings.

After the first T & G, I reported to the pilot that we had flames coming down the right side of the fuselage. At the same time, people in the city of Merced, California, were calling the Castle AFB flight control tower to report that a plane flying around was on fire.

The Instructor Pilot brought the plane around and we landed quickly. Just after the drag chute came out and the plane slowed down, the number 10 hydraulic pack blew up and started filling my gunner's compartment full of smoke and flames.

I noticed that the fire trucks were already coming to our aid so I did not jettison my turret fearing they would drive up under it. I went to the gunner's entrance hatch and opened the hatch. As I pushed the hatch open and the rescue people sprayed me with the foam fire retardant.
I could not get the hatch to stay open because my gloves were so soaked with the slippery foam that I could not grip the handle. There was not a stand available to get me out of the plane and a very bereave fireman was standing below and shouting: "Jump Sarge, I will catch you."

I weighed 220 pounds at that time and would probably have been charged with manslaughter if I had jumped down on him. The escape rope was available but I could not keep the hatch open so I could not use it. I got out eventually with the help of a step ladder and the whole crew was taken to the flight surgeon's office.

Once in his office, he individually checked us each over medically and questioned us about how we felt. When he finished, he gave the pilot a quart of Old Methuselah that melted the paper cups and we shared that before leaving his office.

Another version of a Thai Baht bus.

Bye Lao

My first tour of duty at U-Tapao was as a TDY Wing Gunner from Dyess AFB. There was a man there on a year tour that was afraid to leave the base. For 7 months he did not venture past the front gate. One day he decided to "take a chance" and go to one of the small villages located a short distance from the base. He was very nervous and had no knowledge of the language so we ("being kind souls") took him under our wing. We told him that the Baht Bus drivers usually drove too fast and if he thought they were going too fast to just holler "Bye Lao" until he slowed down. Of course, this meant to go faster and it resulted in a wild ride for our ex-buddy.

He made it back safely, no thanks to us, and didn't have any kind words for us later.

The Final Talk

I often think about the last conversation I had with a fellow gunner before Linebacker II.

His name was Walter and we were in the barrack bathroom standing at one of the many sinks. I was shaving with my electric razor and he started asking about the comfort level of using the electric razor over the use of a blade.

Our talk led to a discussion of our sons and he told me that his son would get a rash under his chin and bumps on his face after using a blade. Walter asked if I thought my type of razor would help him avoid that. My answer was that I hoped it would do the job for him because my son used an electric one and had no problems.

I don't know if my suggestion worked for his son or not because he did not make it back from a mission over North Vietnam. It has been over 30 years since that conversation but it is still etched in my mind. I remember Walter very well and of all the talks I had with people during my career, I will never forget that simple conversation. Who could have expected that he and I would never get to discuss our families with each other again?

We all lost many friends during those times.

Put Out the Arc Light, Free the POGS, and Other Personal Memories

Bill Beavers

I asked Sandy, my wife, should I take a "Northern tier" assignment in B-52H models that had no Southeast Asia tours; or the B-52D at Carswell AFB in Ft. Worth, Texas, and regular tours to SEA. That would mean four months over there and then six months back stateside at that time. She said she didn't like the idea of sub-zero temperatures, had relatives near Carswell, and so I decided I should do for real what I had trained to do - drop bombs. So, Carswell was it. It was almost five years there for her personally and a total of 13 months out of those five years there for me personally. That meant A LOT of time on Guam, Okinawa, and Thailand for me. I became a Vietnam vet that never set foot on Vietnamese soil but had a few million air miles flying to, from, and over the country in B-52Ds (the "black bottom beast"). My first crew was Ron Lane (AC), Dale VanTillborg (CP), with Harry Rumzek (RN), Hal Barry as EWO, and Joe Adams (G). That was 1970. Judd Ruth was our Bomb Squadron Commander. I later moved to the 20th BS with Laymond Ford as Ops Officer. My flight records show me with 3.4 hours in FB-111 flight time there. The FB-111 CCTS was just starting up at Carswell and somebody made a wrong transcription – I never set foot near an FB-111.

Random Memories of my Crewdog Days

Having 8-12 rum and cokes around sunrise after returning from a three-hour "monkey killing/palm tree splitting/probable truck park/POL storage/transhipment point" bombing mission flown out of U-T.

Watching the U-T caged (Pete the) python swallow a live chicken.

Hearing about how those "hoodlum Crewdogs" ran the domesticated/refined PCS party OWC out of the Andersen Officer's Open Mess (OOM) because of our vulgar and crude flight suited attendance there.

Watching a movie at the "Washout" outdoor theater on Guam.

We were the white helmet boys for a long time. Heck, why camouflage helmets like TAC, after all, all they had was experience. But SAC had their own rules and regulations and we had to follow them.

"Tent City" on Guam for the B-52G guys.

Lawrence Welk's recording of "Baby Elephant Walk" played on the UHF radio as the BUFF herd taxied for take-off in Linebacker II on Guam.

Personally eating two of the 1/2 chicken box lunches while out over the Gulf of Tonkin before the turn up the Red River on my first Linebacker II mission. It might have been my last meal if we were shot down or captured by the North Vietnamese and there was no reason to let the chicken burn up in a fire ball in the airplane!!!

When are they going to put out the Arc Light?

Dog meat or yesterday's drop dead water buffalo sold as "Kobe Beef" at various locations in Thailand. Just keep the room dark, add lots of seasoning and sauces and it would taste "fine." Just remember to not ask any questions.

Sattahip jewelry stores where Thai gold was sold in "Baht-bracelets" and "necklaces" that were transported to the U.S. and often sold at great mark-up value.

"Papa-San" and "Mama-San" chairs that filled the bomb-bay racks of bombers returning stateside and the cargo-space of stateside returning tankers.

Loss of almost 100% of my own personal hearing in the decibel range of a screaming J-57 engine.

"Straphangars" who would climb on a "safe" (one going to bomb palm trees in South Vietnam) B-52 bombing sortie to get their "combat pay tax exemption" for the month and never participate to relieve anyone on a 12 hour Guam mission. "Just put my name and Social Security Number on the AFTO 781 please." A few were quite welcome and good to have along. Most of them you never saw before the flight and never saw again.

"Charlie Tower." They were a Crewdog's "angel" (patient, calm, knowledgeable, effective and capable) - despite all the crap that was falling down around their ears and alligators that were trying to swallow them whole - all simultaneously.

The Royal Thai Naval Air Force. When their two or three SA-16s or S2Fs aircraft wanted to taxi and take-off or land at U-T - absolutely all B-52 and KC-135 movements were HALTED NOW IMMEDIATELY!!! IMMEDIATELY!!! OK, it is their airfield, but come-on. Putt-Putt-Putt until they took off or were in the chocks. We wouldn't want to offend the King.

I flew four Linebacker II missions off "the Rock" in B-52Ds. Several Linebacker I ("Compression/Press-On" missions) out of U-T just prior to that. A total of four consecutive TDY tours with Mizner (AC), Don Craig (CP), Chuck Bristol (RN) --- the four of us stayed on the same E-57 crew together throughout. We had various EWOs and Guns. We were the "POGs" (Prisoner's of Guam)/the "Just in Case Force"/the bored "Why the Hell Are We Still Here Force?"/etc./etc. Then I went to Squadron Officer's School at Maxwell AFB, AL in May 1973 with the "horde" that got released to go from the Arc Light/Young Tiger force. Then it was back to Carswell and the 20th Bomb Squadron and my last trip to SEA in the B-52D with Jim Platt (AC) who was a cross-trainee out of C-141s, and Jim Sautter (a brand new, 2nd Lt. Nav) and me as RN.

Then it was a move to Minot AFB in 1974 to the 5th Bomb Wing as a B-52H RN (via Castle AFB CCTS for requalification in B-52H). About a year and a half on crew there and PCA to the 91 Strategic Missile Wing as Minuteman III ICBM crew commander (again via Vandenberg SAC crew training).

After not quite seven years at Minot (and seven winters there!!), I made a PCS to Fairchild and the 92nd Bomb Wing as RN in 1980 for about a year. I made Lt. Col. on arrival at Fairchild. Then went to DOX War Plans in Bomb Wing, Base Operations Division Chief. Site Activation Task Force Detachment for deployment of Cruise Missile to Fairchild, back to DOX War Plans as Conventional Weapons Planner. Retired at Fairchild in 1988.

I consider I contributed two good things to my country for 23 years of service (basically all in SAC) --- we do not speak Russian here today and two, helped get our POWs out of North Vietnam.

Vietnam was a war fought exactly backwards. "Strategic Airpower" was used tactically in South Vietnam while TACAIR was used strategically in North Vietnam. And the rolls were not reversed until the very end of the war. Semper Fi and "Hats Off" to the Marine and Navy and USAF fliers who flew missions in the North for many years, especially to those who flew and "lived with" the on again/off again "No bombing North of such-and-such Parallel" lines. And to LBJ - the stress of the War cost him his life. MacNamara? Well, he built Edsels.

The author at U-Tapao.

Fading Memories, Odds and Ends, And Bar Stories

Tommy Towery

The following memories are not detailed enough to demand their own chapters, but they are some to which many Crewdogs can relate; therefore, these tidbits are included in this section.

One of my favorite flight suit patches of all times was not the "Southeast Asia War Games Participant" that many crews loved (but that was a close second). My favorite patch during that period was the one that said "Bullet Shot – The Herd Shot Round the World." Operation Bullet Shot was the program that launched the massive movement of bombers and crews that carried over 8,000 TDY personnel to Guam in about a week's period.

The "dollar bill" game was often used to decide who on my crew bought drinks a lot of times at U-T (U-Tapao). The person who bought drinks the last round of drinks (i.e., the last loser) would not have to play, but would be the one who held the dollar bill during the guessing. We used the last two numbers of the serial number, and each person playing took turns trying not to guess those numbers correctly. The first person would guess a number between 0 and 99. The holder of the bill declared the guess either high or low. If the guess was low, the next person had to guess a number between that number and 99, if it was high, then the next person had to guess a number between zero and that number. The end numbers would shrink the spread between the numbers. The guesses would continue until someone guessed (or was forced to guess, by default) the correct two numbers. That person had to buy the next round.

When Bullet Shot started, I was working on the wing staff at Carswell. I got the call at 3am and was told to report at 8am, packed to be gone for six months. I did not ferry one of the wing's B-52s over, but I was scheduled to fly on a contract DC-8 flown by a civilian flight crew. I was part of a large contingent of wing staff and maintenance personnel, mostly dressed in our olive drab utility uniforms or 1505 khakis. I wore my flight suit and was the only one in the whole group of about 100 that was wearing one. One of the young stewardesses aboard must have had a thing about flyers and uniforms, because she started paying special attention to me from the start of the very long flight. An hour or so after takeoff she came up and started a friendly conversation with me. We were talking about all sorts of various things when she started asking me about my wings and the patches on my flight suit. She saw the colorful patch of the 7th Bomb Wing, which was bright yellow and embroidered with an eagle with a bomb in its talons, along with the Latin words "Mors Ab Alto" printed below it. In her sweetest smiling voice she pointed to it and said, "What a pretty patch. What do those words mean?" I replied to her that it was the wing's motto, "Death From Above." Her charming smile slowly turned to a frown as she digested what I had just said. With a new sense of understanding, she snapped her head back and, after stuttering a few undistinguishable syllables, she shrieked, "That's awful!" Even though it was a small plane and a long ride, I never saw her nor spoke to her again during the rest of the 12-hour trip.

The little black and gold ribbon that was attached to a Seagram V.O. liquor bottle would be attached to your watch band when you had less than 30 days left on your TDY assignment to Guam. It was affectionately called "The Short Ribbon."

A common saying was that a Crewdog came back from Thailand with two things: a Seiko watch and divorce papers.

Besides the Seiko watches, Crewdogs bought many gold "baht" necklaces and four-seasons bracelets in Thailand. A tailor made suit would cost $20, and a hand-made pair of elephant leather shoes would set you back $10. You were measured for the shoes by placing your foot in a notebook and the cobbler used a pencil to draw an outline of your foot on the page. It usually took two days to get either made, but in a rush, you could be sized and fitted one afternoon and pick it up the next morning.

Another favorite purchase in SEA (Southeast Asia) was a small, square flashlight made by Sanyo. The light was one of the first available plug-in rechargeable types, had both red and white bulbs, and was just the right size to fit in the arm pocket of a flight suit. There was a rubber stopper in the end that could be removed and used to unscrew the light bulb when it needed replaced. The light was affectionately called a "Habu Light." Habus were snakes – which seemed to have no relationship at all to a flashlight, but that did not seem to matter.

The "Falcon Code" was crude, rude, and socially unacceptable, but edged its way into everyday conversations of TDY Crewdogs in Guam and Thailand. It was a shortcut numbering system for popular sayings designed for one number to represent a complete sentence or slang term. There are still copies of the entire list floating on the internet, but one number that I remember well was "109" the code short for "Beautiful, just f*ing beautiful," which sometimes reflected a good bomb run. Many people thought that the flightline snack bar in UT was named the "105 Shack" because of the F-105 fighter. No, it was name that because of the "105" Falcon Code, which stood for "Its so f*ing bad, I can't believe it!"

If you went out the back gate at Andersen and drove through the jungle, you'd arrive at Kenny's Steak House. No menu there, the waitress just walked up to your table and asked if you wanted chicken or steak –that was they had and all you could order.

The wildest New Year's Eve party I ever attended was at Andersen AFB, Guam, and it took place on December 31, 1972 during the 11-Day War period. The permanent party decided they just had to have a party, despite the fact that a war was going on and, at first, the word was sent out that no TDY personnel would be allowed into the Officers' Club. That did not set well with the TDY Crewdogs who had been flying bombing missions over Hanoi less than 24 hours earlier; so they went to the party, anyway. During the celebration, someone (who had taken a survival kit out of a Buff) put sea marker die in the swimming pool, fired pen-gun flares out over the golf course, and took a life raft into the middle of the dance floor and pulled the tab right in the middle of the celebrating crowd, which pushed them in all directions as it inflated.

The wing staff at Andersen got really upset when the crews kept painting over the big sign erected on the base that proclaimed "Our mission – Bombs on Target" to "Our Mission – Boobs on Target."

They also got upset at the underground movement that kept sticking homemade stickers on the seats of crew busses that pleaded - "Free the POGs" (Prisoners of Guam).

During Bullet Shot, I was moved in and out of living quarters frequently. I stayed in a high-end Hotel cabana on the beach in Aguana, a trailer, a tin semi-round Quonset hut, the BOQ, and in a tent just off the runway where the Buffs were launching almost every hour on the hour. It seems like each time a new group was sent to Guam, the whole housing picture was redrawn.

One night at the Andersen AFB Officers' Club, my pilot was given the duty of escorting a visiting general officer for the evening. My pilot walked up to our radar navigator, who had been sitting at the bar for many drinks, and said "Dave, I want you to meet the General." The pilot saw his career coming to an end when the RN spun around on his bar stool and said in an elevated voice "General? What general? I don't see no f*ing General!"

Speaking of generals, I had an interesting encounter with one in Bomber Op's at Guam one night. It was shortly after I arrived there via Bullet Shot, and I was working in the wing's Penetration Aids office. Since Bomber Op's was a 24/7 operation, I had just finished briefing a crew that was headed out for a late night takeoff. I had put all my paperwork away and was just sitting behind the desk starting to work on the next brief just a few hours away when my peripheral vision detected someone entering the room. I looked up, and to my surprise, I was face to face with a man wearing not the normal flight suits that most of my visitors wore, but instead had on a blue dress Air Force uniform with so many stars on the shoulders that I couldn't count them in my shock. It was none other than the Commander in Charge of the Strategic Air Command (CINCSAC), General Bruce K. Holloway, himself! I stood silent as he smiled and casually asked me how it was going. He moved to the side of the room and sat down in a chair that was out of view from the hall and started into small talk. "Go ahead, sit down!" It was the only order I had ever received from CINSAC in person, and I followed it to the letter. I sat down in my chair. He started asking about when I arrived in Guam, where I came from, and other such questions as that. I really don't know what all we talked about, because I remained somewhere between fright and shock at just being in the room alone with him. I noticed a lot of commotion in the hall outside the room and saw blue suited Captains and Majors were streaking back and forth past my door in both directions like they were participating in a fire drill. The General kept on with his chit-chat conversation, and I sat in my chair as he had instructed me to do. After a few more minutes of small talk, a young looking Captain with a panicked look on his face stuck his head in my door and was about to ask me a question when he looked over and saw CICSAC sitting in the chair in the corner. The Captain's face turned to one of relief, and he let out a big breath as he told the General that he was needed in the briefing room. General Holloway slowly stood up, shook my hand, said it was nice talking to me and "Good luck!" and left. I will never forget that "Four-Star" visit.

Some crews went together and bought old rusted Japanese import cars which could barely run. These were affectionately called "Guam Bombs."

We were ferrying a Buff from Anderson to U-T when we lost aircraft pressurization. Had it been a bombing mission, we would have probably aborted into the Philippines, but since we had no weapons on board, we elected to continue the mission. We flew at 25,000 feet and stayed on oxygen most of the time. When we landed at U-T, the Flight Safety Officer met the plane screaming mad and was going to write our crew up for not aborting into Clark AFB (like he felt we should have done). About that time, the Wing Commander drove up and congratulated my pilot, Denny Scruggs, and the whole crew for taking the initiative to get his badly needed replacement plane to his base. The Wing Commander had our crew named "Safety Crew of the Month," which really upset the Flight Safety Officer who wanted to hang us. Thanks to the upset Flight Safety Officer, our reward for being named "Safety Crew of the Month" was to give up our free time and go to each flight safety meeting for the month and brief the crews on what we did. One afternoon we went down to one of the local embroidery shops and had patches made for the whole crew that proclaimed us

"Scruggo's Safety Screw" complete with an embroidered picture of a safety pin piercing a condom. They'd make a patch for anything if you paid their fees.

The favorite song of the TDY crews that was sung by the Philippine band that played at the Officers' Club at Andersen was their adaptation of the Animals' "We Gotta' Get Out of this Place." Every Crewdog in the club sang along whenever it was sung – every time. I cannot hear that song today without thinking of the Guam Officers' Club.

A favorite song sung at the U-T Officers' Club was "Yell-Row Leave-Er," better known as Christie's "Yellow River" here in the states.

A favorite Thai dish was "kowpot" which was a version of "flied lice" (or "fried rice" as it was more properly called). It was cheap and bountiful.

Other cheap items included jewelry and clothing. Many crews took advantage of the cheap tailors (who came from India) in Thailand and had "party suits" made which were colorful decorated flight suit variants.

There was a 24/7 Roach Coach (commercial food vending truck) that sold 25-cent hot dogs set up on blocks outside of the crew briefing room at Andersen. It was located between several palm trees and affectionately called "Gilligan's Island." Many a Crewdog ate midnight hot dogs before embarking on a bombing mission.

One day, we started leaving our alert aircraft, crawled on the bus, and could not find the radar navigator. Since it was a two-man policy violation for any one person to be inside the ropes of the restricted area alone, two of us were sent back to the plane to check it out. As we got near the hatch we heard a voice quietly, but with panic tones, yelling "Get me out of here!" We had left him alone when he climbed on the ladder to check the serial numbers on the bombs, and the rest of us went back to the truck because we couldn't see him on the ladder in the bomb bay.

Following an emergency four-engine landing we were taxing back to the ramp when the copilot started doing the post-landing checklist and shut down the remaining four good engines, leaving us sitting on the taxi ramp with no power, no radios, and no way to start the engines back up.

I remember my Oklahoma copilot having a fit when we came back from a flight out of U-T and found that the hooch-girls had spent most of the day polishing his cowboy boots, giving them a shiny finish and making them look almost brand new after he had spent years roughing them up to look like authentic cowboy boots.

There is a famous story about someone trying to order a hamburger at the U-T Officers' Club dining room and the waitress telling him "No hab hamburgers – hab cheeseburgers!" Think about it.

Maybe this story is an Urban Legend, maybe it really happened. The story goes that a Crewdog was in the U-T Officers' Club dining room, and he told his Thai waitress that he wanted a piece of blueberry pie. The Thai waitress was hampered by limited English and lack of knowledge about what blueberry pie really was. She went to the kitchen and came back empty handed a minute later and told the fellow, "No hab brueberry pie." The guy had ordered it before and knew that it was available one time, so he asked her again to please go check and bring him a piece. She came back again empty handed and a little flustered, saying a little louder, "No HAB brueberry pie!" At that time, he got up, went over to another table, picked up a saucer that had a piece of blueberry pie on it, and showed it to her, demanding that she go get him a piece right now. She walked back to the kitchen to tell the cook, again. A minute later, she swung open the door that connected the kitchen with the dinning hall and shouted across the dinning room, in a voice that silenced the rest of the customers in the room, "NO HAB F**KING BRUEBERRY PIE!" Perhaps she was only echoing the cook's comments. Who knows if it is true, but it makes a great story.

While I was grounded for kidney stones, I was assigned to work in 43rd Bomber Op's at Andersen. On the last SEA bombing mission launched from Andersen, we copied a card out of a popular magazine and had it printed onto and inserted into the mission materials for the three B-52s in the final cell that was launched. The card read "Your subscription ends with this issue."

I remember sitting on a sextant case for a seat and flying on a Buff from Carswell AFB, TX to Barksdale AFB, LA. The plane had the wheels locked down and wheel locks permanently installed. My ejection seat and most of my EW equipment had been removed to be used as parts in other planes. We flew it at 8,000 feet because that plane was designated a "Crested Dove" aircraft which was a plane that was taxied or towed around a base to confuse the Russian satellites trying to keep track of how many we had and where our B-52 bombers were stationed.

Singha Beer in Thailand was reportedly one of the worst tasting beers brewed in the world. It was featured in Newland and Madam T's. Enough said. Think kerosene.

During one static display at an air show at Carswell, my crew was manning the plane to answer any questions the civilian visitors asked. One little boy walked into the open B-52D bomb bay and looked around in amazement. He said quiet innocently while looking up into the empty space "Where do all the bombs go?" My radar navigator, at the time, leaned down to him, took an imaginary cigar out of his mouth, thumped off the ashes, and with his best imitation of Groucho Marx, said, "Well kid, when I drop them, you never can tell!" The kid never understood the answer but the rest of the crowd that did burst into laughter.

Another static display at a different time was manned by my former pilot. His favorite story was when a little kid looked into the cockpit at all the flight instrument dials, then looked up and stated in amazement, "Man, when you wants to know what time it is, you wants to know what TIME it is!"

On a satellite alert deployment from Carswell AFB, Texas to Sheppard AFB, Texas, I opened my B-4 bag in my quarters and found that my daughter had put a "Sniffles the Mouse" pillow in it when I was not looking. The normal pillows in the alert facility were so flat that I decided to use it for extra support, and it worked very well for sleeping. One morning, my roommate took my flight cap and set it on the head of Sniffles and then sat him back onto my cot. Sniffles was setting there, wearing my cap, when the Base Commander made an unannounced visit to show some of the local civilian politicians the living conditions and the professionalism of the SAC alert force. One step forward and two steps back for his career – thanks to a little mouse.

Think of the story line for "Smoky and the Bandit." When Coors beer was not sold east of the Mississippi River, my crew had flown a B-52 to Tinker AFB, Oklahoma, for a wing modification. While we were waiting in the terminal for a return ride on a tanker, a crewmember from a New York based tanker unit came in and frantically asked if we wanted to buy some cases of Coors beer. We explained that we were from Ft. Worth and could buy all the Coors beer we wanted. He said that they had 110 cases they had to get rid of in an hour. It seems that they gone to a local beer store in their flight suits and bought it all, planning to take it back to New York and make a killing by selling it for twice the price they paid. Another customer in the store asked how they were going to get 110 cases back to New York and keep it cold and they bragged that they were going to take it on a USAF airplane. The civilian called the base commander and asked if that was what his tax money was being used for, to pay for beer to be transported for someone to make a profit. The base commander drove out to the plane, told the tanker crew that they could carry back one case per crewmember and to unload the rest from the plane. He reminded them that they had a scheduled takeoff in one hour, and they were expected to takeoff on time. I don't know what happened to all the beer. I think Smokey won that time.

It was not all fun and games, and there are many deeply depressing memories about this period of my life. The area in which I worked in 43rd Bomber Operations during Linebacker II was also the place where the secure phone which was normally used to contact MACV was located. The whole hectic area became deadly still when the wing's intelligence officer walked into the room carrying a small stack of index cards and needed to use the phone. The cards he carried were the Search and Rescue (SAR) cards with the personal authentication questions and answers used to identify any crew member that went down behind enemy lines. These questions were used to identify each crew member because they were personal, such as "What was your first dog's name?" The answers would only be known by the individuals who had filled out the cards. All work came to a stop as the names and the questions were read into the phone headset, and we learned first- hand of the identities of another crew that had been shot down. It was chilling!

On my second deployment to from Carswell to Guam, the tanker aircraft had a maintenance delay, and all the passengers were told to get off the plane, but not leave the immediate area. I sat on the hot Texas tarmac and vividly remember talking to another B-52 staff officer about going to Guam and Thailand and the bombing. I saw him in passing a few times after we got there and we always spoke but I never had another serious conversation with him. His name was Lt. Col. Keith Heggen, mentioned earlier in this collection. He was one of the staff members aboard one of the B-52s that was shot down over Hanoi. He did not survive that incident, and though I barely knew him, I often think of him and that short conversation we shared sitting under the wing of the tanker in Texas.

In one of the oddities of war, when I was grounded for the kidney stones, I was placed in the Penetration Aids office to work. Even though I was a junior Captain and the job required a senior EWO, I made the move, anyway. There were already two Lt Col's in that office and my move there overstaffed the positions. One of the Lt Col's there was forced back to a position on a regular flying crew. That Lt Col I replaced was the EWO on the Carswell crew that was shot down during the Linebacker II operations. That crew was lucky in a sense that they were the only B-52 crew that was shot down over Hanoi that returned intact. The gunner of that crew had a special patch made that he wore on his flight suit that said "2 ½ Missions Over Hanoi."

A normal EWO training class at Mather AFB, CA consisted of around eight people. Of the members of my EWO class, one classmate was killed in a B-52 training accident at Beale AFB, CA; one was killed on a B-52 training mission out of Guam; one was shot down over Hanoi in a B-52 and became a POW, but later returned alive; and one was forced to bail out over Thailand from a damaged B-52 hit by a SAM over Hanoi. Flying is an inherently risky business. Despite many complaints and misunderstandings by the non-fliers in the Air Force about our "undeserved" monetary benefits, we knew that we did not get "flight pay" as they loved to call it; we got "hazardous duty" pay, which we earned. More than once I was known to ask one of the desk jockeys who was complaining about how much flight pay fliers received how many of his specialty were killed sitting at their desk.

Mission Debrief

Conclusions

Bill Beavers

This is the kind of stuff that Crewdog memories are made of. This is what makes us "we" and not one of "them."

It is the "glue" that binds us for a lifetime and places us together at a point in time when "we were there, did that, and survived the episode."

We can now look back on it, through some "rose-colored" glasses and say, "Thanks God for the good people in my life. Thanks for allowing us to share together our mutual good fortunes and even our not so good fortunes. Thanks for helping me to understand why you placed me here both now and back then."

Events and things happen and material things are involved, but in the end friends are what matters and what endures.

B- 52 Crewdogs in Books and Movies

"Well, boys, I reckon this is it - nuclear combat toe-to-toe with the Roosskies. Now look, boys, I ain't much of a hand at makin' speeches, but I got a pretty fair idea that something doggone important is goin' on back there. And I got a fair idea the kinda personal emotions that some of you fellas may be thinkin'. Heck, I reckon you wouldn't even be human bein's if you didn't have some pretty strong personal feelin's about nuclear combat. I want you to remember one thing, the folks back home is a-countin' on you and by golly, we ain't about to let 'em down. I tell you something else, if this thing turns out to be half as important as I figure it just might be, I'd say that you're all in line for some important promotions and personal citations when this thing's over with. That goes for ever' last one of you regardless of your race, color or your creed. Now let's get this thing on the hump - we got some flyin' to do." – A speech delivered to his crew by Major T.J. "King" Kong in "Dr. Strangelove or How I Learned to Stop Worrying and Love the Bomb", 1964.

Those famous words spoken by Slim Pickins reflected some of the irony in the feelings about Nuclear Combat in America in the Sixties. It was the funny movie about B-52 duty that sometimes seemed a little too real to be a comedy, and sometimes a little too comical to be real. In its own sarcastic way it epitomized some of the feelings about the Crewdog's view of life.

The collective group of writers of this book was asked what movies and books they would recommend to the readers who wanted to further understand the lives of B-52 Crewdogs. Here are the ones that were suggested:

Movies

Strategic Air Command (1955)
Bombers B-52 (1957)
A Gathering of Eagles (1963)
Dr. Strangelove (How I Learned to Stop Worrying and Love the Bomb) (1964)
Failsafe (1964)

Books

Linebacker: The Untold Story of the Air Raids Over North Vietnam. by Karl J. Eschmann. Ivy Books : New York City, 1989. ISBN 0-8041-0374-7. LCCN 89-91317. Published in paperback by Ballantine Books.

The 11 Days of Christmas: America's Last Vietnam Battle. by Marshall L. Michel, III. Encounter Books : San Francisco, 2002. ISBN 1-8935-5424-4 (cloth), 1-8935-5427-9 (pbk.). LC Call Number DS558.8 M532 2001. Dewey Call Number 959.704'348.

Boeing B-52 Stratofortress. by Peter E. Davies and Tony Thornborough (with Tony Cassanova). Crowood : Ramsbury, England, 1998. ISBN1-8612-6113-6.

Where the Buf Fellows Roamed. by James Hooppaw. Red Apple: Gig Harbor, Washington, 1999. ISBN 1-8802-2233-7. LCCN 99-74598. In paperback.

Tall Tail Tales by James Hooppaw. Available through AV8R and is $15.00 plus $2,50 shipping and handling for a total of $17.50. (add $.50 s/h for each additional copy) ISBN 0-9722384-1-7. Contact AV8R
West 13518 Meadowview Lane
Nine Mile Falls, WA 99026

Unchained Eagle: From Prisoner of War to Prisoner of Christ (Hardcover) by Robert G. Certain. ISBN 0882801511. *ETC Publications*. Available at Amazon.com

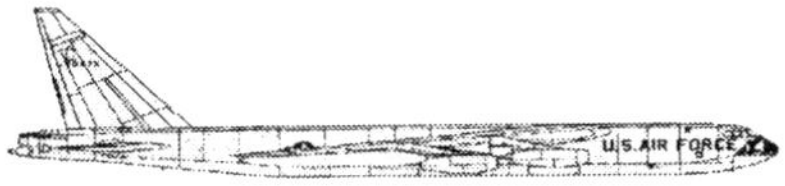

The Contributing Authors

Bill Beavers

Lt. Colonel., USAF (Ret.) graduated ROTC from Fresno State College, California in 1965 with a B.A. in history. He went on active duty to Little Rock AFB, Arkansas in September 1965. He attended basic ICBM school at Sheppard AFB, TX, followed by SAC ICBM crew training at Vandenberg AFB, CA. About three years on Titan II Combat Crew in 308 Strategic Missile Wing/373 Strategic Missile Squadron.

Off to Navigator school at Mather AFB (Sacramento, California) in 1968 (Class 70-05 with [I think] about 60 graduates). Then got married to a young lady I had met at Little Rock (where my initial assignment was as a Titan II ICBM launch officer 1965-68). Then to Navigator-Bombardier school at Matter because I wanted a B-52H ("H," not the older models) model assignment in the southern U.S. (Hell, surprisingly enough, I didn't know there WASN'T any such basing arrangement for the B-52Hs --- they were all in North Dakota, Northern Michigan --- COLD, brrrrrrr country). Should have gone to ECM school instead with my interest and knowledge about it (my avocation for 50 years has been the development of the American military night fighter airplane and the ground and airborne radars that made them possible, 1940-1955). But that is another story.

Walter J. Boyne

Colonel, USAF (Ret.) Former director of the National Air and Space Museum of the Smithsonian Institution, enlisted as a private in the United States Air Force in 1951 and retired in 1974 as a Colonel with more than 5,000 hours in a score of different aircraft, from a Piper Cub to a B-52. For his achievements, in 1998 the National Aeronautic Association named him an Elder Stateman of Aviation.

The Federation Aeronautique Internationale, the international aviation organization of which NAA is a member, honored Boyne with its 1998 Paul Tissandier Diploma.

He has written more than 400 articles on aviation subjects and is one of only a few authors to have had both fiction and nonfiction books on *The New York Times* bestseller lists. His nonfiction books include *The Smithsonian Book of Flight, The Leading Edge, Weapons of Desert Storm,* and *Boeing B-52: A Documentary History;* his fiction books include *The Wild Blue* (with Steven Thompson), *Trophy for Eagles, Eagles at War,* and *Air Force Eagles.* Most recently, he has been instrumental in the development of the Wingspan Air and Space Channel, a 24-hour television network devoted to aviation.

Charles N. Brown

Lt. Colonel., USAF (Ret.) For almost a year prior to the printing of this book, every method available was made to locate and contact Lt. Col. Charles N. Brown. E-mails to his listed addresses were bounced and the letters sent to him were returned as "Moved – No Forwarding Address." Every possible flying and B-52 related association that had a mailing list was contacted to try to locate him. Despite all the efforts, none were successful. It is with that regret that the excerpts have been included without his personal permission despite the granting of the reprint rights obtained from The *Friends Journal*, the quarterly publication of The Air Force Museum Foundation, Inc. We feel that the descriptions and facts he has included in his writings vividly portrayed his message and that any effort to improve upon them would not do the subjects justice. He has done a superior job in his account of the included phases of a Crewdog's life and it is only fit to be reprinted in this collection. We do so with the regret that although we did legally obtain reprint rights, we wish that we could have contacted the author personally as well, and would still like to contact him if someone has a current address.

Robert Glenn Certain

Colonel, USAF (Ret.), was born in Savannah, Georgia in December 1947. He attended Emory University in Atlanta, Georgia, graduating with a BA degree in History in 1969. A distinguished graduate of the Air Force ROTC program there, he was commissioned a second lieutenant and became a navigator-bombardier, flying B-52 aircraft. At the end of his second tour of duty in Southeast Asia, he was shot down over Hanoi, North Vietnam in December 1972, and held captive until the end of March 1973. As a result of his active military service, he was awarded the Distinguished Flying Cross, Bronze Star, two Purple Hearts, five Air Medals, the Prisoner of War Medal, Meritorious Service Medal, Commendation Medal, Vietnam Service Medal, Republic of Vietnam Cross of Gallantry, and Vietnam Campaign Medal.

When stationed in Blytheville, Arkansas, Robert met and married Robbie Lee Wade. He was ordained deacon in the Episcopal Church in July 1975 and priest in April 1976. Chaplain Certain served as an Air Force Chaplain at Andrews AFB in Washington before leaving active duty for the reserves.

As a chaplain, he has continued to be active in the Air Force Reserves, and attained the rank of Colonel. Early in his reserve career, he was attached as a unit reservist to Kelly AFB in San Antonio, Texas. Upon moving to Mississippi, he was initially attached as an individual mobilization augmentee to Keesler AFB, and then transferred back to his original operational base in Blytheville, Arkansas. With the move to Texas in 1989, Chaplain Certain was reassigned from Eaker Air Force Base to Lackland Air Force Base in San Antonio, Texas for Reserve duty. In 1991, he was assigned to the Command Chaplain's office, Strategic Air Command, to coordinate the training of reserve chaplains attached to the command. When SAC was deactivated in June 1992, Chaplain Certain was reassigned to the Command Chaplain's office, Air Mobility Command. In July 1994 he was selected to attend the Air War College in residence at Maxwell AFB in Montgomery, Alabama. While there he wrote a critique of reserve chaplain employment during Desert Shield/Desert Storm for the Chief of Staff of the US Air Force as well as a paper concerning the proper role of military chaplains as ethical advisors to commanders and combatants. Following graduation in June 1995, he was reattached to the Staff Chaplain's Office at the United States Air Force Academy, where he coordinated the work of reserve chaplains, CSSPs, and chaplain candidates in support of Basic Cadet Training until his retirement in July 1999.

Mr. Certain is the founding president of Unchained Eagle Memorial Society, which is currently raising funds to construct a major memorial to POWs and MIAs at the Riverside National Cemetery. His autobiography, *Unchained Eagle: From Prisoner of War to Prisoner of Christ* was published in July 2004.

George R Dempsey

Captain, USAF (Honorably Discharged) I earned the Air Medal with 10 oak leaf clusters for my participation in the Southeast Asia bombing campaign. My active duty time was 5 years 4 months and 3 days. Following Navigator School, I attended Nav-Bomb Training. After CCTS at Castle AFB, CA I joined a crew at Dyess AFB, TX as a Navigator in the B-52D. I never became a Radar Navigator. I spent a lifetime in the belly of my Mother Buff. "The Man" would say approximately 2,650 hours with 335 combat missions flown from Guam and Thailand. If you count the time I have spent reliving the experiences, I would say about 47,000 jillion hours! We now lived in Denver and have owned a small sales company, GDI Sales Inc., for the last 23 years. My wife Janice and I have plans to soon become "Professional Recreationalists".

George W. Golding

Colonel, USAF (Ret.) was born in Lake Placid, New York. He graduated from Hudson High School, Hudson, New York and enlisted in the United States Air Force, going to Lackland AFB, Texas, as an Airman Basic. During Basic Training he was accepted into the Aviation Cadet Training Program and transferred to the Pilot Training Class of 61-D attending Preflight Training. He was commissioned a Second Lieutenant following Aviation Cadet Primary Flight training at Spence Air Base, Georgia, and Basic Pilot training at Vance AFB, Oklahoma.

His first assignment as an Officer was a B-47E copilot at McConnell AFB, Kansas, Forbes AFB, Kansas, and Pease AFB, New Hampshire. His next assignment took him to Carswell AFB, Texas, from 1965-68 as a B-52 copilot and aircraft commander. From 1969-1971 he was assigned to Tachikawa AFB, Japan, and Naha AFB, Okinawa, as a C-13OA pilot and instructor pilot. During this period he flew 287 combat missions. In 1971, the Colonel was a Mission Commander for the 834th Air Division, Tan Son Nhut A B, Republic of Vietnam where he planned and flew as the "On Scene Commander" for 22 Commando Vault missions. Colonel Golding next assignment was to McCoy AFB, Florida, during 1972-1973 where he was a B-52D

Instructor Pilot and Quality Control Officer. During this period he flew 115 combat missions over targets in Southeast Asia, including Linebacker I and II Missions over North Vietnam. In 1973 he was the ARC Light Duty Launch Control Officer ("D" CHARLIE) at Andersen Air Force Base, Guam

Reassignment in 1974 took him to the 68th Bomb Wing Seymour Johnson AFB, North Carolina where he started off as the Maintenance Quality Control Officer and then became the Commander of the 68th Field Maintenance Squadron (FMS). He was selected in 1975 to command the 4008th FMS at Barksdale AFB, Louisiana; later becoming the Wing Maintenance Control Officer. In 1977, Colonel Golding was assigned as Director of Logistics Analysis at Headquarters Strategic Air Command, Offutt AFB, Nebraska.

From 1979-1980 he was a student at the Air. War College, Maxwell AFB, Alabama. Upon graduation, he was appointed Chief of Air Warfare Studies and Analysis at the War College. In 1981 he reported to Hill AFB, Utah, to be the Chief of the Resources Management Division, Directorate of Maintenance, Ogden Air Logistics Center.

Colonel Golding became the Vice Commander of the 320th Bombardment Wing in July 1982 and was named the Wing Commander in May 1983. In May 1986 he retired as the Deputy Chief of Staff for Logistics, Headquarters 15th Air Force, March AFB, CA after 27 years of service. He returned to work as a Logistics Analyst and Logistics Manager with the Boeing Military Airplane Company in Wichita, Kansas for four years before retiring with a medical disability. As an active duty spouse, he has been active in working as a volunteer in Wing Protocol, OSC Boards, ALS and NCO school speaker, Family Services and base Chapel activities.

His educational degrees include a Bachelor of Arts Degree from Louisiana Technical University in 1977 and a Master's Degree in Public Administration from Auburn University in 1980. He completed Air Force Squadron Officer School in 1965, Air Command and Staff College in 1973, and Air War College in 1980.

Colonel Golding is a command pilot with over 7,800 hours of flying time. His decorations include the Legion of Merit, Distinguished Flying Cross with V & oak leaf cluster, Bronze Star/ V, Purple Heart,

Meritorious Service Medal with two oak leaf clusters, Air Medal with nine oak leaf clusters, Air Force Commendation Medal with three oak leaf clusters, Republic of Vietnam Cross of Gallantry with Palm, and other service awards and decorations.

He is married to Colonel Susan J. Golding, who is the Commander of the 75th Maintenance Group at Hill AFB, Utah.

Gordon Hallgren

Major, USAF (Ret.) was commissioned through OTS in December 1968 and served initially as a Public Information Officer. After completing Undergraduate Navigator Training and Electronic Warfare Officer Training in 1972 he volunteered to fly AC-130 Gunships in Southeast Asia. Upon returning home he was assigned to B-52s at Fairchild AFB, WA until 1976 when he went to RC-135s at Offutt AFB NE. He served there as flight and classroom instructor and Standardization/Evaluation officer.

In 1981 he was selected to serve on the SAC Inspector General staff. His final assignment was to the staff of the newly created joint United States Central Command at MacDill AFB FL, where he retired in 1988.

Gary Henley

Colonel, USAF, (Ret.) was commissioned in the Air Force in 1973 through the Texas A&M University Reserve Officer Training Corps program. His operational flight experience includes the B-52D, RS-135S (Cobra Ball), and RC-135X (Cobra Eye) aircraft. He has served as a flight instructor/evaluator, crew commander, flight commander, operations officer, and squadron commander. He has an extensive background in electronic and information warfare (EW & IW) training, systems engineering, and defensive systems flight-testing. He has served in various staff positions at Wing, Center, MAJCOM, and Agency levels. He earned technical specialty badges in both navigation (electronic warfare) and intelligence career fields. Some of his recent duties included Assistant Deputy Chief of the Central Security Service; Chief, M04, National Security Agency; and Chairman of the National Emitter Intelligence Subcommittee (under the National Signals Intelligence (SIGINT) Committee).

He recently retired after 30 years of service in the USAF. His last duty assignment was as vice commander of the 67th Information Operations Wing, Lackland Air Force Base, TX. Currently, he is the Director for the San Antonio office, Information Technology Center, of Syracuse Research Corporation and serves in the capacity of Senior Technical Consultant to the Technical SIGINT Airborne Program Office at the National Security Agency.

His awards include the Defense Superior Service Medal, Legion of Merit Medal, Meritorious Service Medal with four oak leaf clusters, Air Medal with one oak leaf cluster, Air Force Commendation Medal, Air Force Achievement Medal and Combat Readiness Medal with one oak leaf cluster.

James D. Hooppaw

Colonel, USAF (Ret.) is a retired command pilot with 26 years in SAC. He flew the Boeing B-47E and the B-52,C,D,E,F,G, and H models. He served as a pilot, instructor, Squadron Commander, and Deputy Commander of Operations of a Bomb Wing. He served three ARC LIGHT tours in Southeast Asia and received the normal complement of decorations with gongs that everyone else received. Any success he has enjoyed is mainly the result of "working with and for a great group of people."

Arthur Craig Mizner

Major, USAF (Ret.) is a B-52 Combat Veteran Command Instructor Pilot of Vietnam (1969 - 1973) Arc Light, Linebacker I and II missions, and a B-47/B-52 Veteran Pilot of the Cold War ground and airborne nuclear alerts. Craig entered USAF active duty in 1954 as an Aviation Cadet (class 56Q) and was commissioned a 2nd Lieutenant and Jet Fighter Pilot on 28 June 1956 and retired 1 January 1977 with 264 TAC and SAC combat missions, 9206.7 hours in the B-52 and over 11,000 military flight hours. In addition, Craig has over 2,000 hours as a commercial pilot. Craig joined General Dynamics/Lockheed Martin in 1979 and was the F-16 Pilot Flight Manual Manager for the first 17 years. Since than, Craig using his computer skills works as an Advance Technology System Integrator Aerospace Staff Engineer in support of the F-16 *Fighting Falcon*. Craig has BS in Industrial Technology from Texas A&M – Commerce, TX December 1978 and an AA in Social Science from Tarrant Country College, TX May 1998. Craig was inducted in the *Phi Theta Kappa* International Honor Society on 20 April 1998. In 2000, Craig started night school working on a Computer Science degree. For Craig's involvement in the bombing of Hanoi, NVN for the return of all POWs, read "The Eleven Days of Christmas: America's Last Vietnam Battle" by Marshall L. Michel. In May 2002, Craig and his co-pilot Donald Allen Craig were interviewed and video taped for a TV feature on the bombing of Hanoi. The TV feature should air in the near future. See URL for details: http://www.teleproductiongroup.com/12_72-main.html Be sure to navigate to all areas. "Those Who Lived it." The first picture you will see is below. Look at the fifth interview.

Karl D. (Ned) Nedela

SMsgt, USAF (Ret.). Was born and raised in Crete, Nebraska. After high school, attended Doane College in Crete and played 2 years of football. In 1951, I knew I was about to be drafted (not by the NFL) so I enlisted in the Air Force.

After a year in Electronics Courses at Lowry AFB, I attended B-36 Gunnery School in Denver. I went through B-36 Transition at Carswell AFB. Our group was sent to Loring AFB and I served as an instructor gunner for 3 years. In 1956, I was sent to B-52 Gunnery School at Lowry AFB and then to Castle AFB. Returned to Loring and our Squadron was sent to Biggs AFB, El Paso, in the Summer of 1959. Reassigned to Dyess AFB, Abilene, Texas in 1966. In June of 1970, I went for seven months to U-TAPAO as a TDY Wing Gunner.

I graduated from Mc Murry College in 1972, thanks to the Bootstrap Program. Graduated in August and in September went as a crew gunner to U-TAPAO. I returned in April of 1973 and retired in July of 1973. We moved to Killeen, Texas where I taught and coached for 23 years.

I have been married to my wonderful wife, Elizabeth, for 53 years and have 2 children and 3 grand children. I have fond memories of my Air Force career

Vincent (Vince) H. Osborne

Colonel, USAF (Ret.) Born in Nebraska in 1943. Attended high school and Colorado State College (now University of Northern

Colorado) in Greeley Colorado graduating in 1965 with a BA in Education and a commission in the Air Force through ROTC. First assigned to pilot training at Enid Oklahoma. Moved on to navigation and electronic warfare schools at Mather AFB, Sacramento Calif. First operational assignment was at March AFB, Calif. as a crewdog, then "Stan Board" with a couple of short tours to South East Asia (SEA). On to Carswell as an EW instructor and more tours to SEA. Eventually flew 123 actual missions over SEA and earned a Distinguished Flying Cross and the Air Medal with 8 oak leaf clusters. While at Carswell I married Becky (smartest thing I ever did) Subsequent assignments:
SAC Headquarters-Operational Aircraft Test Officer; AF EW School-Flight Commander & Operations Officer; Supreme HQ Allied Powers Europe-Long Term Defense Program Officer. Unified Space Command HQ-Operations & Communications Pentagon; Acquisition officer for Under Secretary of the AF for Acquisition then Under Secretary of Defense for Acquisition

Retired in Aug 1992 and moved to Texas where I am working in the public sector helping people get off welfare and into work.

Denny Scruggs

Colonel, USAF (Ret.) is from Charlotte, NC. Denny graduated from the Univ. of North Carolina at Chapel Hill, after attending the US Military Academy from 1963 to 1964. He entered the USAF in 1967 as a Distinguished Graduate of the Reserve Officers Training Corps. He holds a master's degree in aerospace science and technology from Embry-Riddle University. He's also a graduate of Squadron Officers School, Air Command and Staff College and the Air War College.

While in the Air Force he earned the aeronautical rating of Command Pilot and additionally earned the rating of Master Aircraft Maintenance Officer. He is married to the former Jody Nichols of Seaford, Long Island, N.Y. Denny and Jody have 3 children and 6 grandchildren.

Denny's military service began 1 July 1963, when he took the oath of allegiance on the Plain at West Point. During Denny's almost 3 decades of military service the family lived in 12 different states and 2 foreign countries. He is a Vietnam veteran with 1,062 hours of combat flying time. Col. Scruggs flew the B-52G, B-52F, B-52D, KC-135A, HU-16 and O-1E aircraft. He commanded the 379th Heavy

Bombardment Wing, Wurtsmith AFB, Oscoda, MI, during Desert Storm, 1990-91. He was the last Headquarters SAC Logistics Division Chief for MSET in 1992.

During his career in SAC he was assigned to Robins AFB, Goose AB, Labrador, McCoy AFB, Carswell AFB, Andersen AFB, Griffiss AFB, Minot AFB, Offutt AFB (twice), Plattsburgh AFB, and Wurtsmith AFB. Denny and his crews, both KC-135 and B-52D, pulled extensive TDYs at Clark AFB, Philippine Islands, Kadena AB, Okinawa, Andersen AFB, Guam and U-Tapao AB, Thailand. He retired in September 1992, four months after the Strategic Air Command was closed.

Wayne Shimet

Major,USAF (Ret.) College of San Mateo, AA in General Studies, San Francisco State College, BA in International Business Central Michigan, MBM Personnel Manaagement.

He is married to Nancy Shimet who is a legal secretary. He has one Son Troy Shimet who is a Registered Nurse working at Florida Hospital, Orlando, FL

His assignments include: AF Electronic Warfare Officer, B-52's 51st Bomb Squadron, Seymour Johnson, AFB, NC; Electronic Warfare Officer, RC-135's, 343rd Strategic Reconnaissance Squadron, Offutt AFB NE; Exercise Plans Officer, HQ SAC, Offutt AFB, NE; Faculty / Course Manager: Combined Air Warfare Course, Maxell AFB, AL; Faculty / Course Manager: Senior Tactical Battle Commander's Course, Hurlburt Field, FL

Following his military service he worked at the Okaloosa County Property Appraiser's Office as an appraiser. He is currently retired.

Tommy Towery

Major, USAF, (Ret.) earned a B.S. degree in Journalism in 1968 from Memphis State University where he also earned a commission as a 2nd Lt. through the AFROTC program. Attended Navigator Training and Electronic Warfare Officer Training at Mather AFB, CA. Following B-52 CCTS training in B-52F models at Castle AFB, CA he was assigned to the 20th Bomb Squadron at Carswell AFB, TX flying

B-52C and B-52D models. He served as a 7th Bomb Wing Combat Intelligence briefer and a Penetration Aids staff officer while grounded from flying status for kidney stones. He was deployed for six months to Guam as part of "Operation Bullet Shot" and was assigned to 8th Air Force Bomber Operations as an Arc Light mission planner. Three month after returning from his first deployment he was sent on a second six-month TDY to Guam where he worked as an Arc Light planner in the 43rd Bomb Wing Bomber Operations. It was during this assignment that he flew on B-52 combat missions as a staff officer and helped plan Linebacker II missions. Upon return to full flight duty status he was assigned to a B-52D crew at Carswell AFB, TX. Shortly thereafter he deployed to Guam and Thailand with his first crew and later progressed from crew duty to instructor to standboard duty on Crew S-1. In 1976 he and his crew received the Mathis Trophy, awarded to the top bomber unit based on combined results in bombing and navigation in the SAC Bomb-Nav Competition.

Following his B-52 assignment he was transferred to the RC-135 program at Offutt AFB, NE and flew operation reconnaissance missions from forward operating bases. In 1983 he was assigned to a four-year tour as an Electronic Warfare Intelligence officer and briefer at RAF Mildenhall, UK. Upon his return to the states he spent his last year of active duty as an Electronic Warfare Intelligence officer with the 55th Strategic Reconnaissance Wing at Offutt AFB, NE. During his career he logged over 2,000 hours in B-52s and over 5,000 hours total flight time. His decorations include the Meritorious Service Medal with oak leaf cluster, the Air Medal with eight oak leaf clusters, the Armed Forces Expeditionary Service Medal, the Republic of Vietnam Campaign Medal and the South Vietnam Cross of Gallantry with Palm. He currently is employed as a computer specialist at the University of Memphis and lives in Memphis, TN with his wife Sue who not only graciously allows him to devote many hours to his writing hobby but also serves as his proofreader.

Lynn W. Wakefield

Colonel, USAF, (Ret.) has a MS in Procurement and Contract Management, Webster University and a BS, in Biology. He has held positions as a staff plans and operations officer and as an Electronic Warfare Officer on B-52D/G and RC-135 aircraft. During his career he logged 2,800 hours total flying time. From June 1999 to June 2001, he was Director of Operations at the Air Intelligence Agency (AIA).

From 1997-1999 he served as Commander, 544 Intelligence Group (AIA) at Peterson AFB, Colorado. From 1995 to 1996 he served as Chief, Denver Support Team, National Security Agency. He has also served as Chief of the Requirements and Technology Division, AIA, at Kelly AFB, TX. During his Air Force career he commanded four consecutive Signals Intelligence units: 6993 Electronic Security Squadron, Kelly AFB, TX; 6975 Electronic Security Squadron (Provisional), Riyadh AB, Kingdom of Saudi Arabia; 6913 Electronic Security Squadron, Augsburg City, Fed. Rep. of Germany; Detachment 1, 6903 Electronic Security Group, Osan AB, Rep. of Korea

Since October 2003 he has held the position of Senior Intelligence Specialist, GG-13, Ops Plans Division (AIA/DOX) HQ, Air Intelligence Agency (AIA) San Antonio, TX.

Gerald Wickline

Colonel, USAF, (Ret.) was born in Chillicothe, OH in 1933. He started flying in the USAF via the Aviation Cadet program in 1953. His first assignment was as a navigator on RC-121's at McClellan AFB, CA. He left there and attended Pilot Training in 1958 and was assigned as a B-52 Pilot at Travis AFB CA from 1960 to 1968. His next assignment was as a B-58 Pilot, at Grissom AFB, IN from 1968 to 1971. He served as a Forward Air Controller flying 0-2's in Vietnam in 1971. From 1971 to 1973 he went back to B-52s at Dyess AFB, TX. He retired from the Air Force at Dyess in 1973.

After his retirement Jerry sold Real Estate in Sacramento, CA from 1973 to 1984. He then owned and operated a 44' Commercial Fishing Boat off the coasts of California and Oregon 1984 to 1992. He retired again in 1992 and moved to Palm City, Florida. He has made numerous cruises on the East Coast of the US and the Bahamian Islands aboard his 40' Cabin Cruiser from 1992 to 2002. He currently lives in Palm City, FL. enjoying his hobbies of boating, scuba diving, and flying radio control airplanes.

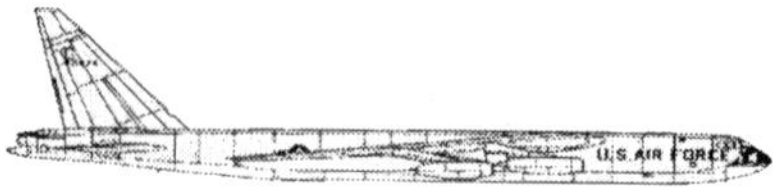